INFORMATION RESOURCES SERIES

Guide to Basic Information Sources
in
CHEMISTRY

Workbook in Chemical Literature and Information Retrieval by Arthur Antony, 1980; developed especially to accompany this *Guide*. Approximately 110 pages; may be obtained from Jeffrey Norton Publishers, 145 East 49th St., New York, N.Y. 10017. *Instructor's Manual* also available for course use.

INFORMATION RESOURCES SERIES
Produced by Jeffrey Norton Publishers
145 East 49th St., New York, N.Y. 10017

Doyle—Guide to Basic Information Sources in English Literature

Mount—Guide to Basic Information Sources in Engineering

Muehsam—Guide to Basic Information Sources in the Visual Arts

Other titles in preparation

Guide to Basic Information Sources in
CHEMISTRY

by
Arthur Antony
Sciences-Engineering Library
University of California, Santa Barbara

A Halsted Press Book
by Jeffrey Norton Publishers, Inc.

John Wiley & Sons, Inc.

New York   Chichester   Brisbane   Toronto

This guide is dedicated to my family, Jo Ann and David.

Library of Congress Cataloging in Publication Data

Antony, Arthur.
    Guide to basic information sources in chemistry.

    (Information resources series)
    1.   Chemical literature.   I.   Title.
QD8.5.A57   1979   540'.7   79-330
ISBN 0-470-26587-6

Distributed by Halsted Press,
A Division of John Wiley & Sons, Inc., New York

Preface and Acknowledgments

This guide is intended to help those who need to find information in chemistry. It is primarily intended for the student of chemistry from college freshman through graduate level—that is, the selections for inclusion have been made and the annotations written primarily with that audience in mind. However, I anticipate that this guide will also be of use to librarians and library school students, and in fact, some items were selected with *that* audience in mind. Indeed, if I have been at all successful in meeting the goals I set for myself in preparing this guide, it should be useful to teachers, researchers, technicians, and all who seek information in chemistry. I apologize to the reader for the extent to which I have failed to meet these ambitions.

This guide has evolved from a similar one that I prepared in 1974 dealing with the chemistry reference materials in the Sciences Engineering Library of the University of California, Santa Barbara, California. That guide, in turn, was an updated version of an earlier guide that had been prepared by Kathy Jackson. Although the present publication is an entirely new work, I owe a debt of gratitude to Ms. Jackson (whom I have never met) for the help that her guide gave me in selecting and organizing the material. I also wish to express my appreciation to Robert Dikeman of Louisiana State University, who taught me a great deal about the literature of science, and to my librarian colleagues in Santa Barbara: Bob Sivers, Virginia Weiser, Norma Claussen, and especially Alfred Hodina. I also acknowledge with gratitude the advice given to me by David Guttman, the University of California, Santa Barbara, media librarian. Without his advice, Chapter 20 could not have been written.

Last, but not least, I gratefully acknowledge the efforts of my wife, Jo Ann, who typed the manuscript of this book, and in the course of doing that, made valuable improvements in the grammar, spelling, and style of writing.

Arthur Antony

Contents

1
Introduction

The vast record of chemical knowledge is the raw material for this book. More specifically, our concern is with the techniques and tools for finding chemical information. The science of chemistry is a consequence of the research that has been reported in the primary literature, mainly in the form of journal articles. In principle, nearly all the information that anyone seeks in chemistry can be gotten from the primary literature. That is why the reference sources, such as bibliographies, indexes, and abstracts, that point to primary documents and their contents are so important. One particular such reference source, *Chemical Abstracts*, provides such extensive coverage of twentieth century chemical literature that an entire chapter of this guide (Chapter 3) is devoted to it.

However, time is a precious commodity for all of us. That surely is true for the researchers, students, teachers, technicians, engineers, and librarians who seek chemical information. Information has been extracted, and often evaluated, in the preparation of many valuable secondary publications. The information seeker may be looking for a single fact or a table of related data. There are, indeed, reference works to satisfy many (but not all) such information needs. For the most part, these reference works are derived from the primary literature, but on occasion they report new information themselves. Some of the most important of these resources are listed in Chapters 12 and 13. The index to this guide can be consulted to help choose one or more reference works that deal with a particular topic. However, the index is not intended to be a comprehensive key, and the user should be aware that the index does not indicate all the properties treated by the reference works listed in this guide.

Other than language dictionaries (Chapter 11), this guide is limited to works in Roman alphabet languages, with emphasis on those in English. Emphasis is also placed on pure chemistry, but some reference materials are included in related areas such as biochemistry and chemical engineering, as well as in applications to environmental concerns, industry, etc.

2
Guides to the Literature

The discipline of chemistry has a tradition of literature guides to help the student, researcher, librarian, or other user negotiate his or her way to needed information. These guides are often especially useful to the non-chemist, for the literature of chemistry is used by people in a great variety of fields. A literature guide directs the reader to significant published sources of information, and may also help identify referral agencies and other non-book resources.

Most of the guides listed here include discussions of important reference tools, such as dictionaries, indexes and abstracts, and encyclopedias. Some concentrate on the various forms of the literature, with special emphasis on the primary sources: Periodicals, patents, governmental and other reports, dissertations and theses, and, in a few cases, manufacturers' trade publications. While some guides are more comprehensive than others, and each has some special features, for many purposes the most appropriate guide will be the one most recently published.

In addition to the guides to chemistry, guides to the general science literature, as well as to other related fields, will be of some use to the researcher or student of chemistry. These are not listed here, and the reader is referred to standard sources such as *Guide to Reference Books*, 9th ed., by Eugene P. Sheehy (Chicago, American Library Association, 1976) or *Guide to Reference Material*, 3rd ed., by A.J. Walford (London, The Library Association, 1973) for the titles of guides in other subject areas.

Periodicals and annuals may from time to time carry articles which may serve as literature guides. The *Journal of Chemical Information and Computer Science* (previously the *Journal of Chemical Documentation* until 1975) carries many articles dealing with various aspects of the chemical literature.

2.1   American Chemical Society. Advances in Chemistry Series. Washington, 1950-.

● The proceedings of the national meetings of the American Chemical Society are not published as units in their entirety. However, the Advances in Chemistry Series includes papers from symposia that were presented at these meetings on a wide range of subjects. Of particular interest to the literature chemist are the following:

2.1.1 No. 4. *Searching the Chemical Literature.* (1951). 184 pp.

● All of the papers deal with *searching* for information. For the most part this issue has been superseded by No. 30 discussed below. Unique to No. 4 is a short paper by Dr. Matthew W. Miller describing the *FIAT Review of German Science*—a massive review of the scientific research that was conducted in Germany during the years of World War II.

2.1.2 No. 16. *A Key to Pharmaceutical and Medicinal Chemistry Literature.* (1956). 243 pp.

2.1.3 No. 30. *Searching the Chemical Literature.* (1961). 326 pp.

● Articles written by members of the *Chemical Abstracts* staff, especially those by E.J. Crane and Leonard T. Capell, are useful guides to that publication through the fifth decennial period. G. Malcolm Dyson offers a guide to searching the literature from between 1750 and 1875 and presents a list of obsolete journals from the nineteenth century. Irlene Roemer Stephens lists sources for identifying dissertations and theses. A list of Russian journals available in translation is included in a rather comprehensive article by Mordecai Hoseh on the Soviet literature. There are several authoritative papers dealing with U.S. and foreign patents.

2.1.4 No. 46. *Patents for Chemical Inventions.* (1964). 117 pp.

● Although the articles in this volume are not intended to serve as literature guides, they provide valuable background information about U.S. and foreign patents.

2.1.5 No. 78. *Literature of Chemical Technology.* (1968). 732 pp.

- Each paper deals with a particular chemical industry, and each contains a very extensive bibliography. An updated version of this very useful source would be most welcome.

2.2   Bottle, R.T. (ed.) *The Use of Chemical Literature.* 2d ed. Hamden, CT: Archon Books, 1969. 294 pp.

- Each of the seventeen chapters is separately authored. Some of the chapters relate to forms of the literature (e.g., primary sources, abstracts, translations), others to subject matter (e.g., nuclear chemistry, polymer science, history and biography of chemistry). There is a particularly useful chapter by T.C. Owen and R.M.W. Rickett on "Beilstein." Bottle has himself contributed a perceptive and comprehensive chapter on "The Use of Standard Tables of Physical Data and Other Physico-Chemical Literature."

2.3   Brown, Russell and G.A. Campbell. *How to Find Out About the Chemical Industry.* Oxford: Pergamon, 1969. 219 pp.

- The British emphasis is particularly noticeable for some of the chapters, such as Chapter 7 on Patents, and Chapter 9 on Safety Measures. Chapter 6 on the Specialized Industries does not provide as extensive an account as the ACS Advances in Chemistry No. 78 (Ref. 2.1.5), but Brown and Campbell do emphasize reference works. Illustrative pages from a number of reference works are reproduced in various chapters.

2.4   Bruce, M.I. Adv. Organomet. Chem., *10:* 273-346 (1972) "Organo-Transition Metal Chemistry—A Guide to the Literature 1950-1970."

- Supplemented by *Adv. Organomet. Chem.,* 11: 447-68 (1973) "Organo-Transition Metal Chemistry—Literature 1971" and by *Adv. Organomet. Chem., 12:* 380-404 (1974) "Organo-Transition Metal Chemistry—Literature 1972." Textbooks, review publications and reference material, including abstracts and indexes, are discussed. Extensive lists of primary and review articles are also included.

2.5   Burman, Charles Raymond. *How to Find Out in Chemistry.* Oxford: Pergamon, 1966. 226 pp.

- This guide is not comprehensive, and perhaps will be most suitable for the information needs of undergraduate and technical school students. Chapter 1 serves as a guide to vocational literature, mainly for Britain, but also for the U.S. Chapter 11, dealing with Societies, is an important feature.

2.6   *Chemical Information Systems.* Edited by Janet E. Ash and Ernest Hyde. Chichester: Ellis Horwood, 1974. 309 pp.

- Emphasis is on computer-based systems that allow retrieval *via* molecular structures and substructures. Linear notation systems, connection tables, and related topics are treated. A few other topics, such as the patent literature, are also included. The individual chapters are written by well-known authorities.

2.7   Crane, E.J., Austin M. Patterson, and Eleanor B. Marr. *A Guide to the Literature of Chemistry.* 2d ed. New York: Wiley, 1957. 397 pp.

- This book is an exemplary subject literature guide in its very thorough coverage of the literature of chemistry at the time it was published. It cannot, of course, provide adequate guidance to the literature of chemistry today, but on those occasions when information about the older literature is needed, it is one of the best sources available. If a guide of this scope were published today, it would probably require several volumes. The extensive treatment of abstract journals (p. 123-157) includes listings by subject of many specialized journals carrying abstracts. For a thorough search of the early literature, particularly on aspects of chemical technology, it might be necessary to consult some of these sources. Another noteworthy feature of this guide is the long table of symbols and abbreviations that constitutes Appendix 2.

2.8   Davis, Charles H., and James E. Rush. *Information Retrieval and Documentation in Chemistry.* Westport, CT: Greenwood Press, 1974. 284 pp.

- The first six chapters deal with general information retrieval principles, with many examples chosen from the field of chemistry. The last three chapters, which physically constitute slightly more than half the book, are predicated upon the high relevance of molecular structures to the information needs of chemistry. The reader must be warned that even though the book was published in 1974, the *Chemical Abstracts* nomenclature discussed in Chapter 7 is relevant to the period before 1972—i.e., before the many changes that were introduced in the 9th collective period.

2.9   Dyson, G. Malcolm. *A Short Guide to Chemical Literature*. 2d ed. London: Longmans, Green, 1958. 157 pp.

• A particularly valuable feature of this guide is Appendix III, which relates volume numbers to publication years for 63 major chemical periodicals.

2.10   Hancock, J.E.II. J. Chem Educ., *45*: 193-99, 260-66, 336-39 (1968). "An Introduction to the Literature of Organic Chemistry."

• The author states that these articles are "intended to give some guidelines to the newcomer faced with the necessity of tracking down information," and that applications of organic chemistry are not covered.

2.11   Houghton, Bernard. *Technical Information Sources. A Guide to Patent Specifications, Standards, and Technical Reports Literature*. 2d ed. Hamden, CT: Linnet Books, 1972.

• British in emphasis. However, the patent systems of a number of other countries, including the U.S., are given brief coverage. There is also a good multinational treatment of standards.

2.12   Hyslop, Marjorie R. *A Brief Guide to Sources of Metals Information*. Washington: Information Resources Press, 1973. 180 pp.

• This book is concerned with sources of information, both published and unpublished on the production, fabrication, treatment, finishing, properties, and applications of metals. In the directory section, specific information can be found about appropriate libraries, associations, government agencies, information centers, etc. The bibliographic information in the first part of the book is keyed to entries in the directory section.

2.13   International Council of Scientific Unions. The Committee on Data for Science and Technology. *International Compendium of Numerical Data Projects*. New York: Springer-Verlag, 1969. 295 pp.

• The numerical data projects of the National Bureau of Standards, the International Union of Pure and Applied Chemistry, and other major international, national governmental, and private organizations, are described, along with the publications that have resulted from their efforts. Major data compilations such as the Landolt Börnstein Tables (Ref. 12.10 and Ref. 12.11) are treated in detail. Hence, this is a very important guide to published numerical data.

2.14 Mellon, M.G. *Chemical Publications. Their Nature and Use.* 4th ed. New York: McGraw-Hill, 1965. 324 pp.

• Mellon's book, intended to be used by students, is a clear and readable account. Especially useful to readers today are his concise accounts of "Beilstein" (Ref. 14.1) and "Gmelin" (Ref. 14.5).

2.15 Mellon, Melvin Guy, and Ruth T. Power. Library Trends, *15:* 836-46 (1966/67). "Chemistry."

• This short bibliographic essay is written for the librarian.

2.16 Moore, John W., and Elizabeth A. Moore. J. Chem. Educ., *52:*288-95 (1975); *53:*240-43 (1976).

• "Resources in Environmental Chemistry." Part I: An annotated bibliography of energy and energy-related topics. Part II: An annotated bibliography of water, life and health, and population problems.

2.17 Smith, J.D., and D.R.M. Walton. Adv. Organomet. Chem., *13:* 453-542 (1975). "The Organometallic Chemistry of the Main Group Elements—A Guide to the Literature."

• This was prepared as a companion to the literature guides by Bruce (Ref. 2.4) and has a similar format.

2.18 Smith, Julian F., and W.G. Brombacher. *Guide to Instrumentation Literature.* Washington: U.S. Government Printing Office, 1965. 220 pp.

• National Bureau of Standards Miscellaneous Publication 271. Supersedes Circular 567. The bibliographic entries, unannotated for the most part, are listed in a classified arrangement based first on format, then within each format, on broad subject. There are indexes by narrow subject and by author, including many corporate authors.

2.19 Woodburn, Henry M. J. Chem Educ., *49*: 689-96 (1972). "Retrieval and Use of the Literature of Inorganic Chemistry."

• Woodburn's essay, intended for the student or instructor in a chemistry literature course, describes techniques for efficient literature retrieval in the field of inorganic chemistry.

2.20 Woodburn, Henry M. *Using the Chemical Literature. A Practical Guide.* New York: Marcel Dekker, 1974. 302 pp.

• Woodburn's approach is that of a chemist instructing chemistry students in the use of the literature of chemistry. Hence, the emphasis is on practical reference works, particularly those useful for laboratory research. Detailed treatments are provided not only for "Beilstein (Ref. 14.1) and "Gmelin" (Ref. 14.5), but also for the Sadtler Spectra (see Chapter 13), "Theilheimer" (Ref. 4.21), the *Methoden der Organischen Chemie* (Ref. 15.2), and other important reference resources that are usually not treated in such detail in guides.

3

Chemical
Abstracts

3.1 American Chemical Society. Chemical Abstracts Service. *Chemical Abstracts*. Columbus, OH: 1907-. Weekly.

Chemists are fortunate in having this comprehensive key to the primary literature, and much of the secondary literature of their field. Issues of *Chemical Abstracts* appear weekly. The odd-numbered issues treat organic chemistry, biochemistry, and chemical history, education and documentation. The even-numbered issues treat inorganic, physical, analytical and polymer chemistry, and chemical engineering. Each issue contains concise, informative abstracts of journal articles, patents, published conference papers, and technical reports, and announcements of dissertations and books along with the complete bibliographic information needed to locate the original documents. There are four indexes at the end of each individual issue: author, keyword, patent number, and patent concordance.

The keyword indexes in the issues of *Chemical Abstracts* are based on author terminology and are not governed by the vocabulary control that presently applies to the General Subject and Chemical Substance Indexes (discussed later in this section).

The numerical patent index is arranged first by country, and then by patent number within each country. Many industrial chemical inventions are patented in more than one country. It is the policy of Chemical Abstracts Service to provide an entry and full abstract only for the first patent they receive for a given invention. The Patent Concordance associates the various patent numbers (from different countries) for the same invention with each other, and with its abstract.

Volume indexes are issued at the end of each volume. At present, one volume covers a six month period, but prior to 1962, each volume covered a year. The volume indexes are cumulated every five years (previously every ten years) into collective indexes. For example, the Ninth Collective Index covers 1972-1976 (ninth collective period). The tenth collective period began in 1977.

The indexing policies for *Chemical Abstracts* have changed over the years. It is useful to first concentrate on present policies, and then look at the past. Few changes have been made between the ninth and tenth collective period, but considerable differences exist between the ninth collective period and earlier periods.

An Index Guide, included as part of the Ninth Collective Index, provides references from many terms that are not used to those that are in the General Subject and Chemical Substance Indexes (discussed below). However, it does not list all the terms that *are* used, nor is there any attempt to include all names that a chemical substance might appear under in the literature. In addition to the "see" references, there are many "see also" references and scope notes in the Index Guide. Since very few cross references are included in the indexes themselves, it is often essential to consult the Index Guide in order to search *Chemical Abstracts*.

The Index Guide also contains useful descriptive information about the indexes, and a hierarchical listing of general subjects. Specific chemical substances are not included in that hierarchical list. The summary of nomenclature rules that is included in the Index Guide is discussed elsewhere (Ref. 9.2).

The Index Guide that accompanies the Ninth Collective Index supersedes the one which was issued at the beginning of the ninth collective period and its supplements. A separate Index Guide for the tenth collective period was published in 1977, and it too will be supplemented, and eventually superseded at the end of the tenth collective period. The Eighth Collective Index also has its own Index Guide, but prior to the eighth collective period, there were no Index Guides.

A short description of each of the volume indexes follows: 3.1.1 Author Index: Personal and corporate authors, patentees and patent assignees are included. Cross references are given from the secondary authors to the first-named author. 3.1.2 Chemical Substance Index: Specific elements, compounds, alloys, minerals, etc. are indexed here. The substance names are based upon *Chemical Abstracts'* highly structured nomenclature rules. Prior to 1972 many non-systematic names were used in *Chemical Abstracts*. Now, however, most of these have been replaced by names which accord with the system. There are still a few exceptions—e.g., a small number of simple compounds such as *acetic acid, phenol* and *urea;* the common amino acids; and numerous natural products with complex stereochemistry.

The Index Guide provides cross references from common names previously used to those now used. However, this is done only for the parent compounds, in most cases. The user may thus have to interpret the cross reference for a related compound in order to generate the name for the specific compound he is looking for. An overview of *Chemical Abstracts* nomenclature is provided in the Index Guide. More details can be found in the *Name Selection Manual* (Ref. 3.4) and in some of the works described in Chapter 9 (Nomenclature).

If, for a given substance, there are a large number of entries, heading subdivisions will be used in order to provide smaller groups. The subdivisions (not all of which need apply in any specific case) will appear in the order: analysis, biological studies, occurrence, preparation, properties, reactions, uses and miscellaneous. There may follow additional subdivisions for certain types of derivatives—for example, if the compound is a carboxylic acid, its esters will be found under the subdivision "esters." As this further subdivision depends upon the type of compound, it is advisable to consult the Index Guide, especially Section 10B of Appendix II.

Names in the Chemical Substance Index are inverted. Thus 1,3-dichlorobenzene is entered as "benzene, 1,3-dichloro." The substituted compounds are listed after all of the entries for the parent compound (including those derivatives that are treated under subdivisions), with the parent compound being represented by a long dash.

benzene

.

.

.

—,1,3-dichloro-

The Chemical Abstracts Service has instituted a system for assigning a unique registry number to each chemical substance. The Chemical Abstracts Registry Number is presented for each substance in the Chemical Substance Index in square brackets, and should not be confused with abstract numbers (which are unbracketed).

In general, documents are indexed in greater depth for the Chemical Substance and General Subject Indexes than they are for the keyword indexes that appear at the ends of the issues. However, it should also be noted that for most chemical substances only one name is used in the

Chemical Substance Index even though the substance may go under many different names. There are a number of exceptions to this. For example, coordination compounds with all but the simplest of ligands will have entries under the central metal and also under the ligand or ligands of interest. However, in general it is essential for the user to be sure he is using the appropriate nomenclature in order to find a substance in this index. Alternative approaches, such as the use of the Formula Index discussed below, often prove more effective than a direct attack on the Chemical Substance Index.

3.1.3 General Subject Index: Prior to 1972, chemical substances and other subjects were all entered in a single alphabet in the subject indexes. Now the index is divided, with all entries that are not for specific chemical substances handled in the General Subject Index—for example, entries for compound classes. In general, compound class entries will be for documents that are about the class as a whole or a significant part of it. Hence, it is essential to look into the Chemical Substance Index for papers dealing with specific members of a class. Subdivisions are also used in the General Subject Index. For compound classes, the subdivisions resemble those already described for heavily posted specific substances.

It is very useful to consult the Index Guide for "see" and "see also" references and for scope notes before using the General Subject Index.

3.1.4 Formula Index: The Hill system is used. That is, each empirical formula is written with carbon first, hydrogen second, and all other elements in alphabetical order if carbon is present, and with all elements in alphabetical order if carbon is not present. The Hill formulas and their arrangement in the index is perhaps best illustrated by an example:

$AsBr_4$	CAl_4O_4	C_2H_3
AsH_3	$CHCl_3$	ClO_3
$As_2B_8H_8S$	CH_2Cl_2	Cl_4Zr
B_4H_9	CNS	Zr
B_4H_{10}	$C_2F_3IO_4$	

The Formula Index can be used to help locate the appropriate name for a compound. Usually, especially for organic compounds, there will be many isomers with the same formula. Note that polymers will be included under the formulas of their monomeric parents, and that some metal salts will also be under the formulas of their parent compounds.

The information in the Formula Index is less complete than in the Chemical Substance Index, in that the Formula Index gives no information about the aspect under which a compound is treated. Abstract numbers usually are included in the Formula Index, but for heavily posted compounds, the user is referred to the Chemical Substance Index for the abstract numbers.

3.1.5 Ring System Index: Over three fourths of the compounds in the Chemical Substance Index contain one or more rings. The Ring System Index offers a systematic approach to the nomenclature of ring compounds based on the number and sizes of the rings and their elemental composition. Only the names of the skeletal rings systems are given in this index; hence, it must be used in conjunction with the Chemical Substance Index.

3.1.6 Numerical Patent Index:

3.1.7 Patent Concordance: Both of these indexes were discussed earlier in this section because they also appear in the individual Chemical Abstracts issues. For these two indexes, the volume indexes can be looked upon as cumulations of the issue indexes.

Literature guides discussed in Chapter 2, especially the one by Crane, Patterson and Marr (Ref. 2.7) include detailed discussion of Chemical Abstracts prior to the ninth collective period. Table 3.1 presents an overview of the indexes to Chemical Abstracts from 1907 to 1971.

A given substance may have had several different names in the indexes, over the years. "Aspirin," for instance, is in the 1907 Subject Index but by 1927, there is, instead a cross reference "aspirin see acetylsalicylic acid." In 1967, the cross reference reads "aspirin see salicylic acid acetate." In the Chemical Substance Index of 1972, there is no entry at all for aspirin, acetylsalicylic acid, or salicylic acid acetate. The Index Guide, however, refers the user from aspirin to benzoic acid, 2-(acetyloxy)-. In the 1977 Chemical Substance Index, there is a cross reference from salicylic acid acetate but not from the other two synonyms.

There has been an enormous increase in the number of chemical substances entered into the indexes with each succeeding decennial or collective index. The majority of compounds in the ninth collective Chemical Substance Index will be unique to that collective index. In the earlier years, a less systematic nomenclature was more acceptable. To some degree, the Index Guide may give guidance to earlier names that were used for specific substances by presenting those names in parentheses at the entries for the currently used names—e.g., "Benzoic acid,

Table 3.1

	Subject	Author	Formula	Ring Systems	Numerical Patent	Patent Concordance
1st Decennial 1907-16	X	X			***	
2nd Decennial 1917-26	X	X	*	**	***	
3rd Decennial 1927-36	X	X	*	**	***	
4th Decennial 1937-46	X	X	*	**	X	
5th Decennial 1947-56	X	X	X	**	X	
6th Collective 1957-61	X	X	X	**	X	
7th Collective 1962-66	X	X	X	**	X	X
8th Collective 1967-71	X	X	X	X	X	X

* In 1951, the American Chemical Society issued a *Collective Formula Index to Chemical Abstracts* covering 1920-1946.

** Ring indexes appeared at the beginning of the Subject Indexes for the 2nd Decennial through Sixth Collective Indexes. For the period of the 7th Collective Index, some guidance is provided by the Ring Index and its supplements (Ref. 3.8).

*** See *Patent Index to Chemical Abstracts* by the Science-Technology Group of the Special Library Association (Ref. 3.9).

2-(acetyloxy)- (*salicylic acid acetate*)." Italics are used for the name in parentheses to distinguish it from a homograph definition which would also be in parentheses—e.g., "seal (animal)" or "seals (mechanical)." Eighth Collective Index names for parent compounds can also be gotten from the Parent Compound Handbook (Ref. 3.7).

The introductory sections of the earlier collective or decennial subject indexes usually contain some guidelines for the indexes and very useful listings of organic groups and radicals. The introduction to the indexes of Volume 66 contains an outline of the nomenclature rules for the eighth collective period. A fairly thorough treatment of the nomenclature rules for the seventh collective period may be found at the beginning of the subject index to Volume 56, and was also issued separately under the title *The Naming and Indexing of Chemical Compounds from Chemical Abstracts* (Chemical Abstracts Service, 1962).

A number of publications are available which deal with *Chemical Abstracts*. Most, but not all, have actually been produced by Chemical Abstracts Service. Some of these are listed below. The abbreviations 8CI, 9CI, etc., are used for Eighth Collective Index, Ninth Collective Index, etc.

3.2 American Chemical Society. Chemical Abstracts Service. *CAS Printed Access Tools: A Workbook.* Columbus, OH: 1977. 270 pp.

- This is a very effective authoritative teaching tool in searching the printed version of *Chemical Abstracts*.

3.3 American Chemical Society. Chemical Abstracts Service. *Chemical Abstracts Service Source Index.* (See Ref. 6.1)

3.4 American Chemical Society. Chemical Abstracts Service. *Chemical Substance Name Selection Manual for the Ninth Collective Period (1972-1976).* Columbus, OH: 1973. 2 vol.

- Chemical Abstracts nomenclature principles are explicated in some detail in this work. Students, chemistry librarians, and others who are going to use the *Chemical Abstracts* Chemical Substance Index a great deal might consider reading all 215 pages of Section A in which spelling, punctuation, and other conventions are explained, and the basic principles of additive, conjunctive, multi-

plicative, radicofunctional, replacement, substitutive, and sub-
tractive nomenclature, and their applications in *Chemical Abstracts*, are described.

Section B deals with the nomenclature of molecular skeletons,
and Section C (a particularly important section) with names for
substituent groups. In the nearly 500 pages of Section D, the more
important compound classes are discussed in alphabetical order.
Numerous well-chosen examples aid the user in understanding the
rules. Special classes of substances, such as carbohydrates, alloys,
coordination compounds, and vitamins, which require specialized rules, are treated in the second part of the work. The
alphabetical index to the entire work is extensive. Another very
useful feature, right before the index, is the illustrative list of
substituent prefixes, which includes in one alphabet prefixes
which are used in *Chemical Abstracts* and cross references from
some common prefixes which are not used.

3.5 American Chemical Society. Chemical Abstracts Service. *Registry Handbook: Number Section*. Columbus, OH: 1974-1975. 7
vol.

• The *Registry Handbook* lists all Registry Numbers for compounds which were registered between 1965 and 1971, and is kept
up to date by annual supplements. It must be noted that some
numbers were omitted from their proper places in the numerical
sequence, and are listed as additions at the end of Volume 7. Each
number is accompanied by the formula and the accepted *Chemical Abstracts* name. The names may be based on the nomenclature policies of the eighth collective period, but if the ninth
collective period name was available at the time the Handbook
was prepared (1974-1975), this is the name that appears. Hence,
for substances which appear often in the literature the name will
be that from the ninth collective period, but it may not be for less
common substances. The supplements, of course, include only
current (i.e. ninth and tenth collective period) names. Since there
have been some number changes, the user is encouraged to
consult the latest Registry Number Update (which cumulates all
preceding Updates), in addition to the Handbook and its supplements.

Registry numbers are now included in quite a few primary
publications, such as the *Journal of Organic Chemistry*, and
secondary sources, such as the *Merck Index* (Ref. 10.31). The
Registry Handbook can be used to translate those numbers into
Chemical Abstracts names.

3.6 American Chemical Society. Chemical Abstracts Service. *Subject Coverage and Arrangement of Abstracts by Sections in Chemical Abstracts.* Columbus, OH: 1974. 212 pp, plus Appendix.

• *Chemical Abstracts* is divided into eighty sections, according to broad subjects. Some users choose to browse sections of interest as a means of keeping up to date. It is also possible to include section and subsection numbers in computer searches for which one or more facets consist of very broad subjects. For either of these purposes, this publication should be consulted, for it explains the nature of the coverage as well as the topics specifically excluded for each of the eighty sections. The subsection breakdown, useful in computer searches, is also given. Changes in the assignment of section numbers have been made, and such changes may need to be taken into account for retrospective computer searches. A chart displaying these changes since 1967 is appended. This publication is included as part of the *Specifications Manual for Computer-Readable Files in Standard Format*, published by Chemical Abstracts Service, or it may be purchased separately. An earlier edition, copyrighted in 1971, provides the same information prior to the changes that were made in 1974.

3.7 *Parent Compound Handbook.* Columbus, OH: Chemical Abstracts Service, 1976-. Biannual.

• This is a very useful nomenclature aid for the Chemical Substance Indexes, especially, but not exclusively, for compounds which contain ring systems. It is divided into two parts: the Parent Compound File (PCF) and the Index of Parent Compounds. It should be noted that most of the entries in the Chemical Substance Indexes are based on the names of parent compounds.

The PCF is divided into four sections, one for each of four different kinds of parent compound. Cage parents are mostly boron-containing systems, but also include some metallocenes and related systems. Acyclic stereo-parents are those non-cyclic parent compounds whose names imply stereochemistry—including, for example, a-amino acids and carbohydrates. Cyclic stereoparents are ring-containing parent compounds whose names imply stereochemistry. All other ring-containing parents are called ring parents, and these constitute the great majority of entries in the PCF. The parent compounds themselves need not exist. If one of more derivatives of a parent have been reported, that parent is a candidate for the PCF. Each entry in the PCF includes a structural formula of the parent compound, with appropriate locant numbering. The CAS Registry Number, the current (9 CI and 10 CI) name, the 8 CI name (if different from the current name), the molecular formula, and (usually) the Wiswesser line notation of each parent compound is also given. Within each of the four sections, the entries are arranged by Parent Compound Identifiers—alphabetical codes which are assigned sequentially and have no structural significance. These codes allow the various parts of the Index of Parent Compounds to be used to search for entries in the PCF.

The indexes by Wiswesser line notation, molecular formula, and CAS Registry Number require no further elaboration. There is an index by parent name, and it is noteworthy that some of the names have been permuted, so that some related parent compound names can be identified. Two indexes relate specifically to ring systems. The Ring Analysis Index is organized like the Ring System Index (Ref. 3.1.5) in *Chemical Abstracts*. For those ring systems in which there is more than one ring, the component rings each serve as an entry in the Ring Substructure Index. This last will be particularly useful, as it makes possible the identification of entries for a given ring system when it is embedded in larger ring systems.

CAS Registry Numbers obtained from the *Parent Compound Handbook* may be used in computer searches of files derived from CASIA (Ref. 5.4). However, the derivatives of such parent compounds will not automatically be searched for in this way. The *Parent Compound Handbook* may also be used as a source for *Chemical Abstracts* nomenclature for parent compounds. This information may serve as a basis for a manual search of the Chemical Substance Indexes for the parent compound and its derivatives. The parent name may also be the basis for a computer search on CHEMNAME (discussed with CASIA, Ref. 5.4) to get registry numbers of derivatives. These last two approaches might not turn up all of the derivatives of a given parent compound because of priority rules in *Chemical Abstracts* nomenclature.

3.8   Patterson, Austin M., Leonard T. Capell, and Donald F. Walker. *The Ring Index; A List of Ring Systems Used in Organic Chemistry.* 2d ed. Washington: American Chemical Society, 1960. 1425 pp.

• Supplements were issued in 1963, 1964, and 1965. For most purposes, this is superseded by the *Parent Compound Handbook*

(Ref. 3.7). Over 14,000 ring systems are described in the *Ring Index* and its supplements. Arrangement is based on number, size, and skeletal composition of the rings. Because the nomenclature is that which was used in earlier years, and because references to primary sources, and sometimes to secondary sources such as Beilstein or *Chemical Abstracts* are given, there still are occasions when the *Ring Index* may be consulted. The primary reference is ordinarily the "earliest one stating the structure with certainty."

3.9 Special Libraries Association. Science-Technology Group. *Patent Index to Chemical Abstracts 1907 1936*. Ann Arbor: J.W. Edwards, 1944. 479 pp.

• An authorized facsimile of the original publication is available from Xerox University Microfilms, Ann Arbor, MI. U.S. chemical patents are first, then those of other countries, alphabetical by country. Within each country, the arrangement is by patent number, with the *Chemical Abstracts* volume and abstract number given for each one.

4
Abstracts and Indexes Other than Chemical Abstracts

Although *Chemical Abstracts* provides comprehensive coverage of the chemical literature, there are occasions when the user might need or wish to turn to some other indexing source —perhaps because of a specialized point of view, or coverage of very early literature, or because the subject is peripheral to *Chemical Abstracts'* subject coverage. Even the most comprehensive source will occasionally miss a relevant document or index one incompletely.

This chapter is limited to abstracts and indexes which are, or were in the past, issued on a regular basis, and which index the primary literature, although some of them might index secondary literature as well. Indexes to the literature of science in general, important as they are to chemists, are not included, except in a few cases where there are sections specifically for chemistry. *Science Citation Index* deserves mention because often the best way to approach the literature of an area of current research in chemistry is *via* its citation index. If one, or a few, key papers can be identified, *Science Citation Index* can be used to locate papers which have cited them. Abstracts and indexes for other disciplines, such as *Biological Abstracts*, *Physics Abstracts*, and *Current Index to Journals in Education* often lead to articles of interest in some areas of chemistry, or in the applications of chemistry. They are not included here, and the reader is urged to refer to guides in the related disciplines of interest.

There are a number of abstracting publications in specialized areas of chemistry which are primarily intended for current awareness, i.e., to alert the chemist to the most recent publications in his area of interest. Although a few major current awareness tools, such as *Chemical Titles* and *Current Contents*, are included in this chapter, the specialized publications of this sort are excluded. The P.R.M. Science and Technology Agency, 787 High Road, North Finchley, London, which publishes several abstracting journals in specific areas of spectroscopy, chromatography, and related subjects, and Information Retrieval Limited, 1 Falconberg Court, London, which produces abstracts on nucleic acids, carbohydrates, and other compound classes of biological interest, are two important specialized publishers whose products are not listed separately in this chapter.

4.1   American Chemical Society. *Abstracts of Papers*. Washington. 1937 ?-. Irregular.

    ● Abstracts of the papers which were presented at the National American Chemical Society meetings are to be found in this publication. In those cases for which the full papers have been published in journals or other sources, they must be located *via* the author indexes of *Chemical Abstracts* or other abstracting and indexing publications. For recent meetings, it is suggested that the author be contacted directly.

4.2   American Chemical Society. Committee on Professional Training. *Directory of Graduate Research*. Washington. 1951/52-. Biannual.

    ● "Faculties, publications, and doctoral theses in departments or divisions of chemistry, chemical engineering, biochemistry, pharmaceutical and/or medicinal chemistry at universities in the United States and Canada . . ." (from page iii, 1975 edition.) This publication is particularly useful as a guide to the recent publications and current research interests in the graduate departments of North American universities. Only programs which lead to the doctoral degree are included.

4.3   American Society for Testing and Materials. *ASTM Publications*. Philadelphia: 1942-. Annual.

    ● This is annotated bibliography of the Society's current publications, along with a list of out-of-print publications which may be purchased from University Microfilms, Ann Arbor, Michigan.

4.4   *Analytical Abstracts*. London: The Chemical Society, 1954-. Monthly.

- Starting with Volume 11, each issue begins with information on published proceedings of conferences of interest to analytical chemists, and includes the authors and titles of all the papers that were published. Abstracts of papers published in journals and other publications, such as standards, follow. There are annual author and subject indexes, and two cumulative indexes, one covering 1954-1963, and the other covering 1964-1968.

4.5 Chemical Society (London). *Journal.* "Abstracts of Chemical Papers" (1871-1923).

- For over half a century, this major primary journal also served as an important secondary source. Until *Chemical Abstracts* began publication in 1907, this was the most comprehensive set of English language abstracts in pure chemistry. Author and subject indexes to the abstracts were begun in 1872. In 1924 and 1925, the abstracts were published separately as Part 2 of the Journal, and were then superseded by *British Chemical Abstracts*, which is no longer published.

4.6 *Chemical Titles: Current Author and Keyword Indexes from Selected Journals.* Washington: American Chemical Society, 1960-. Biweekly.

- *Chemical Titles*, which is based on the tables of contents of over 700 key journals in chemistry and chemical engineering, tends to be a very up-to-date alerting service. It should be noted that the "keywords" are only those which appear in the titles of the articles, and that many complex words, including some chemical compound names, are fragmented, so that entries are provided for parts of some words. Titles not originally in English are translated into English.

4.7 *Chemisches Zentralblatt.* Berlin: Verlag Chemie. 1830-1969. Weekly.

- Begun in 1830 as *Pharmaceutisches Centralblatt*, this publication has undergone a number of changes in title and publisher. It is especially useful for the pre-1935 literature. A brief but perceptive discussion of *Chemisches Zentralblatt* can be found in Mellon's *Chemical Publications* (Ref. 2.14).

4.8 *Current Abstracts of Chemistry and Index Chemicus.* Philadelphia: Institute for Scientific Information, 1960-. Weekly.

• Prior to 1970, this publication was titled simply *Index Chemicus*, and it is still often referred to in speaking by the older, shorter title. It is one of the most important indexes to the journal literature of organic chemistry.

Coverage is concentrated upon approximately 100 journals, which the publishers estimate to include nearly 90% of all the new organic compounds being currently reported. Articles are included only if they report new compounds (including intermediates), syntheses, or reactions. Excellent graphics illustrating the structures of compounds and the courses of reactions contribute significantly to the value of this publication. Author's abstracts from the original articles are included in the entries. Analytical and other techniques are pointed out by means of a coded "wheel" which has been drawn for each entry. Features such as this "wheel", plus the relative rapidity with which most articles get covered, help make this publication an attractive current awareness aid for researchers. It is also of some value in retrospective searches, mainly because of some of the features of the indexes.

There are separate indexes by journal, molecular formula, keyword, biological activities, instrumental data, new reactions/new syntheses/labeled compounds, author, and corporate location. Quarterly and annual cumulations of these indexes, except for the instrumental data index, are published. The quarterly and annual indexes also feature the Rotaform index, an index of permuted molecular formulas for all compounds except those that do not contain elements other than carbon, hydrogen, nitrogen, or oxygen. Hence, the Rotaform index is a very handy search tool for heteroorganic compounds.

Most of the entries in the molecular formula index are to new compounds (only those with asterisks are not). The keyword index (not limited to article titles) is primarily an index by compound name. Many of the entries are non-systematic names, and often entries are given for only parent compounds. There are also a few entries of other kinds, such as reaction types.

An excellent poster, "How to Use *Current Abstracts of Chemistry and Index Chemicus*," which is more instructional than promotional, is available from the publisher.

4.8.1   *Chemical Substructure Index*. Philadelphia: Institute for Scientific Information, 1966-. Monthly.

• This index of permuted Wiswesser Line Notations of the new compounds reported in *Current Abstracts of Chemistry and*

Index Chemicus, is sold separately. The monthly issues are cumulated annually. The publishers issue the *ISI Chemical Substructure Dictionary* to aid the user who is not familiar with the Wiswesser Line Notation.

The Permuted Wiswesser Line Notation allows the possibility of substructure searching—that is, identifying compounds which have specific molecular structural features in common, and hence, locating the journal articles which describe them.

4.9 *Current Contents: Physical and Chemical Sciences.* Philadelphia: Institute for Scientific Information. 1961.- Weekly.

• *Current Contents: Physical Sciences,* which began in 1961, merged with *Current Contents: Chemical Sciences* to form the present title in 1971. Over 700 journals are covered. The tables of contents are produced, and there are subject and author indexes, as well as an author address directory, in each issue.

4.10 Data Compilation Abstracts. See *Journal of Physical and Chemical Reference Data* (Ref. 12.8).

4.11 *Diffusion and Defect Data.* Aedermannsdorf, Switzerland: Trans Tech, 1974-. Semi-annual.

• Continues *Diffusion Data,* which began publication in 1967. The literature dealing with diffusion and defects in metals, ionic crystals, semiconductors and other solids is surveyed. Numerical data, and on occasion, tables and charts are included. Indexes are by subject and by materials, and a cumulative index to volumes 1-10 has been published.

4.12 France. Centre National de la Recherche Scientifique. *Bulletin Signaletique.* Paris: 1940-.

• *Bulletin Signaletique* is issued in numerous parts, corresponding to different subjects. There have been many changes in its division throughout its history. Those sections currently being published of principal interest in chemistry are: 161. crystallography; 165. atomic, molecular, fluid and plasma physics; 170. chemistry; 320. biochemistry and biophysics; 780. polymers; and 880. engineering and industrial chemistry. Section 170 is of particular interest. There are twelve issues per year with abstracts, and indexes by subject (which include some entries for general subjects as well as specific substances) and by author.

Coverage is international, but, partly because many entries of chemical interest are in other sections, the coverage is not as extensive as in *Chemical Abstracts*. The index names provide an alternative to *Chemical Abstracts* nomenclature which might be useful on occasion. All abstracts and index entries are in French, but titles of English and German papers are given in the original languages.

4.13   *International Catalogue of Scientific Literature, 1901-1914.* Published for the International Council by the Royal Society of London. London: Harrison, 1902-1921. 32 vol.

•   According to the 9th edition of the *Guide to Reference Books* compiled by Eugene P. Sheehy (Chicago: American Library Association, 1976, p. 693) this annual bibliography was the most important one covering all the sciences at the time it was issued. Section D covered chemistry. A reprint was published by the Johnson Reprint Corporation of New York in 1968.

4.14   *Mass Spectrometry Bulletin.* Reading, England: Mass Spectrometry Data Centre, 1966-. Monthly.

•   Each issue is composed of abstracts of journal articles, technical reports, and other publications, with indexes. The subject index has entries for specific subjects, and in addition there is a general index which treats general topics such as "conference sites," "materials," and "natural species." The user may need to look in both places. There are also indexes by compound class, element, and author.

4.15   *Metals Abstracts.* Metals Park, OH: American Society for Metals; and London: The Institute of Metals, 1968-. Monthly.

•   The basic chemistry and physics of metals, as well as their technology, is covered. Each issue has a classified arrangement, and an author index. The author index is cumulated annually, and in addition, there is a detailed annual subject index based on a controlled vocabulary. The *ASM Thesaurus of Metallurgical Terms* published by the American Society for Metals in 1976 is the subject authority for *Metals Abstracts*, and also for the METADEX data base. (Ref. 5.8).

4.15.1   *Alloys Index.* London: The Metals Society; and Metals Park, OH: American Society for Metals, 1974-. Monthly.

- *Alloys Index* may be used independently, to retrieve original documents, or it may be used as an index to the abstracts in *Metals Abstracts* (Ref 4.15). *Metals Abstracts* does not provide the detailed indexing by specific alloy that is found in the *Alloys Index*. In the latter, the entries are in the same form as in the original documents, and hence there is some scatter of entries for the same alloy. The index is based on a classification scheme, and there is an alphabetical Alloys Classification Guide at the beginning, by specific alloy name, to help the user locate the appropriate classification for a given alloy.

4.16 *Molecular Structures and Dimensions.* Edited by Olga Kennard et al. Utrecht: Bohn, Scheltema and Holkema, 1970-. Annual.

- This is a classified bibliography of organic and organometallic crystal structures prepared by the Crystallographic Data Centre of Cambridge University. Volumes 1 and 2 cover 1935-1969, with subsequent volumes covering approximately one year periods. At the end of each volume, there are cumulative indexes by transition metal, formula, and author.

4.17 *Mössbauer Effect Data Index.* Edited by John G. Stevens and Virginia Stevens. New York: IFI/Plenum, 1966-. Annual.

- Volume 1 covers the literature for 1958-1965. Volume 2 and subsequent volumes cover shorter periods. This is not a typical abstracting journal. There is a section for each isotope for which Mössbauer literature appeared in the period covered, and the nuclear level scheme demonstrating the appropriate Mössbauer transitions for that isotope, along with a table of basic data, is provided, as well as the abstracts of the literature. There are additional abstracts by topics that cannot be subsumed under specific elements, grouped into sections labeled analysis, general, instrumental, proposal, review, and theory. Among other features, there is an author index.

4.18 *Organometallic Compounds.* Braintree, Essex, England: R.H. Chandler, 1961-. Biweekly.

- Subtitle: Abstracts of literature and patents relating to compounds which contain at least metal, carbon, and hydrogen atoms.

The odd-numbered issues, "Chemistry, production and use in synthesis," are arranged by groups in the periodical table. The even-numbered issues treat a variety of topics of technological interest. The issues have no indexes, but there is an annual subject index.

4.19   *Physics and Chemistry of Glasses*. Sheffield: Society of Glass Technology. 1960-. Bimonthly.

●   Glass is a material of major importance in chemical laboratory practice. Hence, this specialized source (which is not limited to applications to laboratory apparatus) is perhaps of more interest to chemists than similar publications devoted to other materials. *Physics and Chemistry of Glasses* is a primary journal, with abstracts to other publications at the end of each issue. The annual author and subject indexes are both to the abstracts and to the original articles in the journal itself.

4.20   *Spectrochemical Abstracts*. London: Adam Hilger, 1938-1973. Monthly.

●   Emphasis is on analytical applications of atomic absorption, fluorescence, and emission spectroscopy. The arrangement of abstracts is by broad subject area: substances analyzed, apparatus, methods, basic theory, books and reviews; and each of these areas is further subdivided. Conferences, journals, and occasionally, government and industry reports, are covered. Indexes are by element and by author.

4.21   *Synthetic Methods of Organic Chemistry*. Edited by William Theilheimer. Basel: Karger, 1942/44-. Annual.

●   Originally in German, as *Synthetische Methoden der Organischen Chemie*, English editions are available for volumes 1 and 2, and for volume 5 and all subsequent volumes. The abstracts are arranged according to a classification scheme that emphasizes the bond being formed and the nature of the reaction (e.g., elimination, rearrangement, etc.).
       The subject index has entries for compound classes, reaction types, and other terms. There is a formula index for compound types, and the user should be aware that hydrogen atoms are not included in the formulas. The usual Hill system is *not* followed.

There are cumulative indexes for five volume periods. A detailed discussion of the reference work is given by Woodburn in his guide (Ref. 2.20).

5

Bibliographic Searching By Computer

Many indexes and abstracts are produced with the aid of the computer, with the consequence that machine-readable bibliographic products are also available. Thus there are computer tapes that contain the bibliographic citations, index terms, etc., that can be found in *Chemical Abstracts* and similar publications. We shall refer to such tapes as databases. Some organizations subscribe directly to tapes, and program their computers to allow internal users access to the information on the tapes. A number of search services have been established which purchase or lease tapes from various database producers, and then in turn sell access to the public *via* their search systems. The services that concern us are Bibliographic Retrieval Service (BRS) in Schenectady, New York, Lockheed Retrieval Services (LRS) in Palo Alto, California, the System Development Corporation (SDC) in Santa Monica, California, and the Space Documentation Service—European Space Agency (SDS—ESA) in Frascati, Italy. These four services offer online search capabilities to the general public, and include databases of interest in chemistry. In addition, there are a number of centers which carry out searches on their computer facilities for the general public. These will not be listed here, but many of them are included in the second edition of the *Encyclopedia of Information Systems and Services* edited by Anthony T. Kruzas (Ann Arbor, MI: Edwards Brothers, 1974).

The most important English-language bibliographic databases are listed in this chapter. Each search service which offers these databases has its own program for access, and indeed may, in some cases, modify the database, split it up into parts, etc. As this is a field in which there are new developments nearly every week, much of the information in this chapter will soon be out-of-date. *Computer-Readable Bibliographic Data Bases. A Directory and Data Sourcebook*, compiled and edited by Martha E. Williams and Sandra H. Rouse (Washington: American

Society for Information Science, loose-leaf, updated as needed) is a rather comprehensive list of databases in all subject fields, and includes databases which are readily accessible to the general public, as well as some which are not. Detailed information about each database is provided. A column "Data Bases On Line" is published periodically in *On-Line Review* (Oxford, England: Learned Information, 1977-. Quarterly) which lists the databases available through BRS, LRS, SDC, and *SDS—ESA*.

Computer-readable files in specialized subject areas (such as chemical/biological activity and food and agricultural chemistry) are available from Chemical Abstracts Service but they are not listed here. Note that for those databases which are offered through one or more of the search services mentioned above, the date given in the entry is the earliest date that is available for online searching. There may be earlier tape issues of the database itself obtainable from the database producer. The user manuals of the appropriate search services should be consulted to learn about various special features and also for lists of the user aids to be consulted when designing a search strategy. When a controlled vocabulary has been applied in the indexing (as for CASIA, METADEX, and INSPEC, for example) it is usually very important to consult the appropriate thesaurus before conducting a search.

5.1   CA CONDENSATES. Columbus, OH: Chemical Abstracts Service, 1969-.

•   All of the information found in the weekly issues of *Chemical Abstracts*, except the abstracts themselves, is on these tapes. Note that the controlled vocabulary index entries of the General Subject and Chemical Substance indexes are *not* available on CA CONDENSATES. Chemical formulas (unless they appeared in document titles) and registry numbers are also not included. Hence the limitations that apply to the weekly keyword indexes to *Chemical Abstracts* also apply in the case of this database. BRS, LRS, SDC, and ESA—SDS provide online search capabilities to CA CONDENSATES. LRS and SDC handle this database in conjunction with CASIA (see Ref. 5.4 below). Chemical Abstracts Service has announced that it will cease issuing CA CONDENSATES in 1979. However, it is likely that retrospective searches will be performed on it for many years after that.

5.2   CA PATENT CONCORDANCE. Columbus, OH: Chemical Abstracts Service, 1972-.

- This file, available through LRS, corresponds to the *Chemical Abstracts* Patent Concordance (Ref. 3.1.7).

5.3 CA SEARCH. Columbus, OH: Chemical Abstracts Service, 1978-.

- The information in CASIA (Ref. 5.4) and CA CONDEN-SATES (Ref. 5.1) is combined into one database. At this time, none of the search services have announced that they intend to put this file up, but it appears to be the most likely candidate to replace CA CONDENSATES and CASIA when they cease to be issued.

5.4 CASIA. Columbus, OH: Chemical Abstracts Service, 1972-.

- CA abstract numbers, section numbers, index terms (corresponding to the entries in the General Subject and Chemical Substance Indexes), molecular formulas, and registry numbers are included. As full bibliographic citations are not included, search services must tie CASIA to CA CONDENSATES (Ref. 5.1) in some way. LRS and SDC offer CASIA searching. LRS has created a file called CHEMNAME (file 31) from the chemical substance nomenclature, which contains the systematic names, along with synonyms, registry numbers, and molecular formulas of a sample of over 400,000 compounds and other chemical substances. The sample was taken in such a way that most compounds that are mentioned in at least two documents that were indexed between 1972 and 1976 are included, along with many of the compounds that appeared only once in that period. The CHEMNAME file does not contain any bibliographic information; it is, in effect, an online glossary, and is particularly useful in obtaining registry numbers. LRS's files 3 and 4 (covering 1972-1976 and 1977+ respectively) are merges of CA CONDENSATES and CASIA. However, the only chemical substance information from CASIA included on files 3 and 4 are the registry numbers. Thus the searcher should obtain registry numbers (from CHEMNAME or some other source) in order to search for documents related to specific chemical substances. File 2, covering 1970-1971, is based on CA CONSENSATES only. LRS applies an algorithm to the data entered into files 2,3, 4, and 31, such that chemical names are fragmented into significant chemical parts. It is possible to search on full names or on parts of names, the latter making substructure searching possible. The SDC approach has been rather similar in that an online dictionary file of chemical nomenclature has been created separately from the bibliographic file. Chemical Abstracts Service has announced that it will cease issuing the CASIA database in 1979.

5.5   CLAIMS. Arlington, VA: IFI/Plenum Data Co., 1950-.

• U.S. chemical patents, and some foreign equivalents to U.S. chemical patents, are included. LRS provides bibliographic information in the CLAIMS/CHEM files (files 23 and 24). The method of choice for subject searches on these files is to use codes from the U.S. Patent Classification System. Appropriate classification codes may be obtained online *via* LRS's CLAIMS/CLASS file (file 25).

5.6   ICRS. Philadelphia: Institute for Scientific Information, 1966-.

• This database corresponds with *Current Abstracts of Chemistry and Index Chemicus* (Ref. 4.8) and, dating back to 1970, also incorporates the *Chemical Substructure Index* (Ref. 4.8.1). It is thus possible to search this file using Wiswesser notations (which are particularly valuable for substructure searches) as well as the more conventional index entries. At this time, the database is not available for searching from any of the four services described in this chapter.

5.7   INSPEC. London: Institution of Electrical Engineers, 1964-.

• *Physics Abstracts, Electrical and Electronics Abstracts*, and *Computer and Control Abstracts* are the printed publications that correspond to this database. Because of its thorough coverage of the physics literature, this is a good database to include in searches on topics in quantum chemistry, solid state chemistry, statistical mechanics, etc.

INSPEC may be accessed through BRS, LRS, SDC, or SDS—ESA. Words and phrases from the controlled vocabulary may be found in the *INSPEC Thesaurus* (London: Institution of Electrical Engineers, 1976). Search services may also allow the user to include author names, title words, free index terms, words from the abstracts, etc., in the search.

5.8   METADEX. Metals Park, OH: American Society for Metals. 1966-.

• This database available for searching with either LRS or SDS—ESA, corresponds to *Metals Abstracts* (Ref. 4.15) and *Alloys Index* (Ref. 4.15.1).

5.9 NTIS. Springfield, VA: National Technical Information Service, 1964-.

- The corresponding printed publications are *Government Reports Announcements* and *Government Reports Index* (Ref. 7.4) and hence for all practical purposes can be considered to be limited to the coverage of technical reports. NTIS may be searched *via* BRS, LRS, SDC, or SDS—ESA.

5.10 WPI. London: Derwent Publications, Ltd. 1963-.

- This database available for searching through SDC, is the most comprehensive online service for patents for the period covered. Coverage of all chemistry dates back to 1970; some specific classes of chemicals, such as pharmaceuticals, receive earlier coverage.

6

Periodicals and Lists of Periodicals

The primary record of chemistry as a science is to be found mainly in the myriad of journal articles that have been published in the field. Periodicals devoted to reporting results of research in chemistry began appearing towards the end of the 18th century, and prior to that chemistry research was reported in the more general scientific periodicals. Today, as in the past, much research that is of interest to the chemist is published in non-chemistry journals, and many of the producers of chemistry reference works such as *Chemical Abstracts* scan sources from a broad range of scientific and technological subjects to find information of interest to the chemist.

For nearly three centuries, periodical publication has been increasing with an exponential rate. No attempt has been made to list the chemistry periodicals in this chapter. Instead, this chapter is designed to direct the reader to useful lists of periodicals. Guides in the past, when the literature was somewhat less voluminous, did provide lists of periodical titles. A particularly noteworthy example of this was the extensive listing in Crane, Patterson, and Marr's 1957 guide (Ref. 2.7) which classified the periodicals according to *Chemical Abstracts* sections.

The identification of nearly any serial published today in chemistry can be accomplished with the aid of the sources listed in this chapter, especially Ref. 6.1, but variation in the use of abbreviations may be a source of difficulty. The problem of lack of standardization in serial title abbreviations is less serious with recent publications than it was in the past.

For a very comprehensive search for a serial title, it may be necessary to resort to the *Union List of Serials in Libraries of the United States and Canada* or *New Serial Titles*. The third edition of the Union List was published in 1965 with the cooperation of the Library of Congress. *New Serial Titles* is itself a serial publication, appearing monthly, with cumulations covering various periods of time in the past. The first cumulation covers the period 1950-1970. For very early publications, Bolton's bibliography (Ref. 8.5) which will be discussed in Chapter 8 will be of use.

6.1 American Chemical Society. Chemical Abstracts Service. *Chemical Abstracts Service Source Index*. Columbus: 1975. 2 v.

• The previous lists of serials published by Chemical Abstracts under the same title are superseded by this list, which covers the time period from the inception of *Chemical Abstracts* in 1907 to 1974. Quarterly supplements, with each fourth supplement being a cumulative annual supplement, bring the work up to date. A new cumulative edition of the entire index is to appear every five years. The serials, conference proceedings, and monographs which contained collections of papers which have been abstracted by Chemical Abstracts are listed in *Chemical Abstracts Service Source Index* (often referred to as CASSI). Over 14,000 serial titles are included. Of these, there are fewer than 300 titles that have all papers abstracted. Many periodicals are scanned for relevant articles, and as a result CASSI is a very extensive serials list for all of natural science.

The arrangement in CASSI is alphabetical by *Chemical Abstracts* abbreviation. Very often the *Chemical Abstracts* title differs from the standard library main entry which is in accordance with the Anglo-American Cataloging Rules (AACR). For example, where CASSI has *Journal of the American Chemical Society*, AACR would require *American Chemical Society. Journal*. The title according to AACR format is provided with each entry in CASSI, along with the *Chemical Abstracts* title, but it is the latter that determines the arrangement. Titles are in the original languages (tranliterated for non-Roman alphabets). Translated titles are provided for all but those in French, Spanish, German, or Latin. There is a wealth of useful data for each entry, including the history and frequency of appearance of the publication, and a listing of library holdings from around the world. Through cooperation with three other major secondary services (BIOSIS, Engineering Index, Inc., and the Institute for Scientific Information) CASSI is able to indicate which of these services, as well as Chemical Abstracts Service, covers each of the titles. There are, in fact, a small number of titles in CASSI which are not covered in *Chemical Abstracts*. In the beginning of the first volume of CASSI, some additional valuable information is provided, including addresses of libraries which provide holdings information, a directory of publishers and sales agencies, a listing of the 1,000 most frequently cited journals in chemistry, and bibliographic data and source information for official publications of patent offices.

6.2 *Bibliographic Guide for Editors and Authors.* Washington: American Chemical Society, 1974. 362 pp.

• Section 3, which occupies nearly the entire volume, is a list of about 27,500 serials, in alphabetical order by title. These are titles which are or have been scanned in the production of *Chemical Abstracts, Biological Abstracts, Bio Research Index,* and *Engineering Index.* Bold-faced type is used to indicate the abbreviations of the titles. The only other information provided for each title is the CODEN, and an indication of secondary service coverage.

6.3 Emery, Betty L., and Robert T. Bottle. *Gratis Controlled Circulation Journals for Chemical and Allied Industries. A Directory.* Rochester, NY: Upstate New York Chapter of the Special Libraries Association, 1970. 40 leaves.

• Gratis controlled circulation journals are defined in this directory as those ". . . which are distributed *gratis* to qualified personnel in a specific field . . . and rely on advertising revenue for their income." (p. iii).

6.4 Mason, Penelope C.R. *A Classified Directory of Japanese Periodicals. Engineering and Industrial Chemistry.* London: Aslib, 1972. 160 pp.

• Within each subject, entries are listed alphabetically by romanized Japanese titles. In most cases, a translation of the title into English is also provided. Much additional information is given as well, such as the year of first issue, frequency of publication, and language used. It is worth noting that many Japanese periodicals which are published in English or other languages have Japanese titles. However, some authors cite these journals by their non-Japanese (i.e., translated) titles. The index to this classified directory includes both romanized Japanese titles and translated titles.

6.5   Pflücke, Maximilian and Alice Hawelek. *Periodica Chimica.*
      Berlin: Akademie-Verlag, 1952. 411 pp.

   • This listing of the periodicals scanned to prepare *Chemishes
      Zentralblatt* (Ref. 4.7) was reprinted in 1961. A supplement of 245
      pages was published by Akademie-Verlag in 1962.

7
Access to Primary Publications Other than Journals

While the journal article holds the place of honor among the media of primary publication in chemistry, dissertations, technical reports, and other formats cannot be ignored. Resources treated elsewhere in this guide, especially *Chemical Abstracts*, often provide coverage of the publications discussed in this section. In cases for which the best bibliographic approach is *via* general tools, entries are provided for these tools, but annotations are limited to those aspects that pertain specifically to chemistry, and to the basics essential for effective use.

DISSERTATIONS AND THESES

Doctoral and Master degree students carry out a great deal of the basic chemical research as part of their degree requirements. Their dissertations and theses are thus important sources of primary information. The information in the dissertation or thesis may ultimately be published in journal articles or other formats, but sometimes with less detail, and often with the omission of valuable background information and bibliographies.

The thesis or dissertation, more often than not, appears earlier than the journal articles reporting the same research.

7.1 *Comprehensive Dissertation Index 1861-1972*. Ann Arbor: University Microfilms, 1973. 37 vol.

    ● The first four volumes of the title keyword index are devoted to chemistry. Chemical engineering can be found in Volume 9. The index may be used to locate abstracts in *Dissertation Abstracts* (Ref.7.2), but is not limited to what can be found there and hence is more comprehensive. Supplements keep the work up to date. The corresponding computer-readable tapes have been made available for searching through several search services.

7.2   *Dissertation Abstracts.* Ann Arbor: University Microfilms, 1938-.
      Monthly.

      • The name was changed to *Dissertation Abstracts Interna-
      tional* (DSI) in 1970. Chemistry is treated in Section IVB of each
      issue, along with mathematics, physics, engineering, and other
      pure and applied sciences. The section is divided into broad,
      general categories, such as inorganic chemistry, the dissertation
      author having chosen the category. Author-prepared abstracts
      accompany each entry. *Chemical Abstracts* provides citations to
      appropriate DSI entries, but does not itself publish the abstracts.
      Comprehensive Dissertation Index (Ref. 7.1) may be used as a
      cumulative index to DSI.

7.3   *Masters Theses in the Pure and Applied Sciences Accepted by
      Colleges and Universities of the United States and Canada.* New
      York: Plenum, 1955/56-. Annual.

      • Arrangement is by discipline, and then by universities within
      each discipline. Chemical engineering is in Section 7, chemistry
      and biochemistry in Section 8. There are no indexes.

TECHNICAL REPORTS

Technical reports are not easy to define. In general, they are research
reports that are issued separately (that is, not in journal format) but
which most libraries do not fully catalog. They tend to be short,
although a few occupy several volumes. Many libraries collect technical reports in microfiche format and arrange them by accession
number. It is important to note that technical reports are not usually
subjected to the relatively stringent editorial scrutiny and peer review
that journal articles are. On the other hand, some technical reports are
eventually submitted and published as journal articles, often reduced in
detail. Technical reports tend to emphasize applications rather than
basic research.

7.4   *Government Reports Announcements and Index.* Springfield,
      VA: National Technical Information Service, 1946-. Biweekly.

      • This publication has undergone numerous title changes and
      has not followed a consistent format. The reports issued by the
      National Technical Information Service (NTIS) are covered.

These are reports of research sponsored by the U.S. Federal government, but by no means is all such research included. The biweekly announcements and indexes are issued separately, and there are annual cumulative indexes. The corresponding computerized data base, NTIS, has already been discussed (Ref. 5.9).

CONFERENCE PROCEEDINGS

Every year, a lot of original chemical research is reported at conferences, symposia, workshops, and the like. In many cases, there are no publications specifically generated by a conference, although the research that is reported may subsequently appear in journal articles or other publications. Often, however, the proceedings of a conference are published. Bibliographic control of conference proceedings is difficult because of the variety of publishers and formats involved. Many published conference papers are indexed and abstracted in *Chemical Abstracts*, and therefore *Chemical Abstracts Service Source Index* (Ref. 6.1) may occasionally serve to identify a conference publication.

The papers presented at the national American Chemical Society meetings are not published in their entirety. Sometimes symposia on special subjects which were held at the meeting are published, usually as part of the ACS Symposium Series, or as a volume in *Advances in Chemistry*, both of which are published by the American Chemical Society. Abstracts of all the papers presented at the national ACS meetings are published in the Society's *Abstracts of Papers* (Ref. 4.1).

The first three entries below can be used to identify published proceedings as a whole, and give no information about the individual papers in the proceedings. The fourth entry is an index to the individual papers presented at meetings, and the last entries provide information about upcoming meetings.

7.5 British Library. Lending Division. *Index of Conference Proceedings Received*. Boston Spa: 1964-. Monthly.

• This is perhaps the most comprehensive listing of conference publications available, but the only publication-related information given for each one is its main entry. The listing, which is arranged by keywords from the titles, is mainly of use for locating publications in the British Library's very large collection of conference proceedings. However, the establishment of the main entry is very useful for locating more complete information in a library or union catalog. The code numbers in the Index refer to locations at the British Library. They often correspond to journal or serial titles, and may be looked up in *Current Serials Received by the BLL*, published in London by Her Majesty's Stationery Office.

7.6   *Directory of Published Proceedings. Series SEMT*. Harrison, NY: Inter Dok, 1967/68-. 10 issues per year.

- Series SEMT covers science, engineering, medicine, and technology. Though less comprehensive than the *Index of Conference Proceedings Received* (Ref. 7.5.), this publication gives more complete bibliographic information in each entry, and has subject/sponsor, editor, and location indexes, the last being an index by the place where the conference was held. There are also annual cumulative volumes.

7.7   *Proceedings in Print*. Mattapan, MA: 1964-. Bimonthly.

- No single listing of published conference proceedings is sufficient for complete coverage. The two preceding sources (Refs. 7.5 and 7.6), along with this one, as a group provide more comprehensive coverage than any one alone. Very often, it is known that a conference was held, but there is doubt whether or not proceedings were published, and/or there is little or no knowledge about the format, title, etc. of those proceedings. In such a case, a search through each of these three publications in turn may be in order. *Proceedings in Print* is well indexed, and provides satisfactoy bibliographic detail.

7.8   *Current Programs*. Louisville: Data Courier, 1973-. Monthly.

- This is an index to the papers that were presented at recent major meetings in chemistry and other sciences, and engineering. Ordering information about available or forthcoming publications is usually included.

7.9   *World Meetings: Outside United States and Canada*. New York: Macmillan, 1968-. Quarterly.

- The location, sponsors, deadlines for submission of papers, and other information is provided for meetings up to two years in the future. In a few cases, there is mention of expected publications.

7.10 *World Meetings: United States and Canada.* New York: Macmillan, 1963-. Quarterly.

- Previously titled *Technical Meeting Index.* Identical in format to *World Meetings: Outside United States and Canada* (Ref. 7.9).

PATENTS

Chemical Abstracts is a useful source for locating patents issued since 1907 in industrial countries. In addition to the numerical patent index and the patent concordance, which are discussed in Chapter 3, *Chemical Abstracts* provides subject-related entries in the Chemical Substance, General Subject, Formula and Ring System volume indexes, and the keyword issue indexes, and includes the patentees and patent-assignees in the Author indexes. However, for very thorough patent searches, especially where there are legal or high-cost implications, the professional services of a patent lawyer are recommended.

The patent offices of the various countries issue publications which can be used in patent searches. Needless to say, these are not limited to chemical inventions, as *Chemical Abstracts* is, but they are limited to inventions registered in their country of origin. The U.S. Patent and Trademark Office issues the *Official Gazette-Patents* each week.

Each issue is indexed by patentee and classification number. The *Index to Classification*, which is issued by the U.S. Patent and Trademark Office periodically, can be used to find the appropriate classification numbers for given topics. It is advisable to also consult the *Manual of Classification*, which displays the entire classification scheme. The *Index of Patents* is an annual compilation of the classification number index to the gazette.

8

Bibliographies

A fairly large number of bibliographies of interest in chemistry are published every year. In the Library of Congress classification scheme, chemistry bibliographies are classed between Z5521 and Z5526. Over fifty subjects, from aceto acetic acid to zirconium, and including alchemy, alloys, colloids, deuterium, fermentation, morphine, photochemistry and many others, are listed as subdividing class Z5524, Special Topics. Bibliographies on specialized topics can be located *via* most library catalogs by searching for the most appropriate subject heading, and then looking for the subheading "bibliographies." The library user is reminded that books often contain valuable bibliographies, and that especially for research literature, a recent definitive review, either in the form of a periodical article or a book, may be the best source of a good bibliography.

Chapters 2, 4, 6, and 7 treated publications that are, in effect, bibliographies of special kinds. Many of the reference books that are covered later in this guide, especially some of the encyclopedias, contain useful bibliographies. This chapter provides a short list of important bibliographies which do not fit within the parameters of other chapters within the guide. Emphasis is on recent publications, except where very extensive coverage of the early literature is provided.

A particularly valuable function of one kind of bibliography is to help the user locate data for specific properties of chemical materials. Some of the sources listed in this chapter contain data. A subjective decision to include these publications in this chapter, rather than in Chapter 13 was made depending on what was judged to be the primary utility of the book.

8.1   American Society for Testing and Materials. *Molecular Formula List of Compounds, Names, and References to Published Infrared Spectra*. Philadelphia: 1969. 610 pp.

• Subtitle: An index to 92,000 published infrared spectra. Most of the major collections of infrared spectra, including the Sadtler, API, DMS, and others, are indexed in a single list by molecule formula. Spectra which have been published in journals and other primary sources are also indexed. For a given formula, isomers are distinguished by the names used by the authors. Compounds with unknown molecular formulas are included in a separate section. Supplements are issued periodically. The edition published in 1969 incorporates the information in all the supplements through the thirteenth.

8.2   American Society for Testing and Materials. *Serial Number List of Compound Names and References to Published Infrared Spectra*. Philadelphia: 1969. 762 pp.

• This companion publication to (8.1) is a listing by serial number which provides the author-supplied names, but not the molecular formulas, of the compounds.

8.3   *Bibliography on Flame Spectroscopy. Analytical Applications 1800-1966*. Compiled by R. Mavrodineanu. Washington: U.S. Government Printing Office, 1967. 155 pp.

• U.S. National Bureau of Standards Miscellaneous Publication 281. Books and chapters in books, journal articles, theses, and other publications are in an arrangement classified by subject and format.

8.4   *Binary Fluorides. Free Molecular Structures and Force Fields. A Bibliography (1957-1975)*. Compiled by Donald T. Hawkins, Lawrence S. Bernstein, Warren E. Falconer, and William Klemperer. New York: IFI/Plenum, 1976. 238 pp.

• Fluorine forms binary compounds with nearly all of the elements. Fluorine compounds are of interest in the science of chemistry, and many of them also have technological significance. Therefore, this extensive and well-arranged bibliography, though admitted by the compilers to be incomplete, is a valuable one for the research library. Arrangement is alphabetical by name of the partner elements, and there are author and permuted title indexes.

8.5 Bolton, Henry Carrington. *Selected Bibliography of Chemistry 1492-1892*. Washington: Smithsonian Institution, 1893.

• A bibliography of books in seven sections: (1) Bibliography; (2) Dictionaries; (3) History; (4) Biography; (5) Chemistry, Pure and Applied; (6) Alchemy; (7) Periodicals. Some periodical articles are included in sections 1, 3, and 4. Supplement 1, 1492-1897, and Supplement 2, 1492-1902 have, in addition to the seven sections of the main work, an eighth section on academic dissertations. The entire set was reprinted by Kraus (New York) in 1967.

8.6 *Chem Books*. Basel: Karger Libri, 1968/69-. Annual.

• This is a subject-classified bibliography of new chemistry books and newly announced chemistry journals in the Western languages. There is an author index.

8.7 Francis, Alfred W. *Handbook for Components in Solvent Extraction*. New York: Gordon and Breach, 1972. 534 pp.

• According to the forward by Norman O. Smith, this book makes it possible "to locate all the published literature, up to and including 1969, on any given ternary or quarternary systems involving liquid-liquid equilibria." The bibliography is arranged alphabetically by author, and the entries often include citations to *Chemical Abstracts* and other secondary sources. This is an extensive index by component liquids. Approximately 13,000 chemical systems are covered.

8.8 Hawkins, Donald T. *Physical and Chemical Properties of Water. A Bibliography: 1957-1974*. New York: IFI/Plenum, 1976. 556 pp.

• The bibliography is divided into two parts: Part 1 for 1957-1968, and Part 2 for 1969-1974. Liquid water and ice are covered, and there are some references for steam. The bibliography is designed for the physical scientists, and excludes references to natural waters, industrial uses, and environmental and biochemical aspects.

8.9 *Index to Reviews, Symposia Volumes and Monographs in Organic Chemistry.* Edited by Norman Kharasch, Walter Wolf, and Elaine C.P. Harrison. New York: Pergamon, 1962. 345 pp.

● The period from 1940-1960 is covered, and, in addition, organic reviews which were published in *Chemical Reviews* before 1940 are included. Updates were published in 1964 (260 pages, covering 1961-1962) and 1966 (326 pages, covering 1963-1964), edited by Norman Kharasch and Walter Wolf.

8.10 Richards, W.G., T.E.H. Walker, and R.K. Hinkley. *A Bibliography of Ab Initio Molecular Wave Functions.* Oxford: Clarendon, 1971. 211 pp. plus *Supplement for 1970-1973* by W.G. Richards, T.E.H. Walker, L. Farnell, and P.R. Scott. 1974. 358 pp.

● The compilers of this bibliography characterize themselves as users of molecular wave functions, not as primary producers. Some data, such as internuclear distances, and the energy of specific states, is given. Arrangement is by molecular formula, beginning with diatomic molecules.

8.11 Sherwood, Gertrude B. and Howard J. White, Jr. *A Bibliography of the Russian Reference Data Holdings of the Library of the Office of Standard Reference Data.* Washington: U.S. Government Printing Office, 1974. 16 pp.

● U.S. National Bureau of Standards Technical Note 848. "This library is limited almost entirely to material containing critically evaluated data in the physical sciences and closely related reference material such as annotated bibliographies" (p.1). Information on available translations of these works, or translations in progress, is included. Arrangement is alphabetical by author names, with no subject approach.

8.12 Wichterle, Ivan, Jan Linek, and Eduard Hála. *Vapor-Liquid Equilibrium Data Bibliography.* Amsterdam: Elsevier, 1973. 1053 pp.

● The bibliography which is alphabetical by author, covers the literature from 1900 through 1972. The index which precedes the bibliography is arranged by molecular formulas, and includes binary, ternary, and higher systems.

9
Nomenclature

The importance of nomenclature conventions and standards to the development of modern chemistry should not be underestimated, particularly with regard to the names of chemical compounds. As a consequence, a knowledge of nomenclature principles is often helpful, and sometimes indispensable, in searching for information in reference books. The International Union of Pure and Applied Chemistry (IUPAC) is generally looked to by the international scientific community to formulate and publicize principles of chemical nomenclature to be adhered to in formal publication. Consequently, many of the items listed in this section of the guide have been issued by IUPAC.

IUPAC has, in fact, published definitive rules for the nomenclature of important classes of inorganic and organic compounds, as well as for other entities, such as the symbols of physical constants. The journal *Pure and Applied Chemistry* is the organ IUPAC uses for definitive nomenclature, but some of the more important collections of rules are also available as separate publications. Not all classes of compounds are covered by definitive rules. Provisional rules covering many such classes have been published by IUPAC and other bodies, such as the American Chemical Society. Comprehensive coverage of provisional rules is not attempted in this guide, but those pertaining to some of the more important classes of chemical substance are treated in this section.

Nomenclature rules are intended primarily to provide guidance for formal publication. In some cases, the rules recognize more than one name as correct, and as has already been discussed in Chapter 3, organizations which publish massive indexes, such as Chemical Abstracts Service, must formulate more precise rules. Two important publications dealing with *Chemical Abstracts* nomenclature are included in this chapter.

The correct name of a compound is often very unwieldy for conversations and lectures. Trivial, semi-systematic, and trade names are in common usage, and often appear in print as well as in speech. Some of the dictionaries described in Chapter 10 will be of help in connection with non-systematic names. In addition, in this chapter, mention is made of books *about* nomenclature which may be helpful in understanding and utilizing both systematic and non-systematic nomenclature.

The *Journal of Chemical Education* regularly includes a column "Notes on Nomenclature," by W.C. Fernelius, Kurt Loening, and Roy Adams, which provides informative reviews of nomenclature principles for specific types of compounds.

9.1   *Chemical Substance Name Selection Manual for the Ninth Collective Period (1972-1976).* See Ref. 3.4.

9.2   American Chemical Society. Chemical Abstracts Service. *The Naming and Indexing of Chemical Compounds from Chemical Abstracts During the Ninth Collective Period (1972-1976).* Columbus: Chemical Abstracts Service, 1973.

• This reprint of Section IV of the Index Guide (see Chapter 3) of *Chemical Abstracts*, Volume 76, is available as a separate publication. It is a summary of the rules, but in sufficient detail to permit the naming of most of the compounds that appeared in *Chemical Abstracts* during the ninth collective period. An excellent classified bibliography of sources on chemical nomenclature is included. While there are few changes in the nomenclature for the tenth collective period, there were many differences in nomenclature usage for *Chemical Abstracts* prior to 1972. This is discussed in Chapter 3 in more detail.

9.3   Banks, James E. *Naming Organic Compounds. A Programmed Introduction to Organic Chemistry.* 2d ed. Philadelphia: Saunders, 1976. 309 pp, plus separately paged appendices and index.

• This text, designed for self-instruction, is very useful for learning the basic *Chemical Abstracts* nomenclature of organic compounds, and for contrasting that nomenclature with common usage where appropriate.

9.4   Cahn, Robert Sidney. *An Introduction to Chemical Nomenclature.* 3rd ed. New York: Plenum, 1968. 117 pp.

• The IUPAC rules for some of the more important types of inorganic and organic compounds are explained. There is also a chapter devoted to physicochemical symbols, and one in which important differences between American and British usage are discussed. The book, published simultaneously by Butterworths, in London, is from a British viewpoint, and stresses the practices of the Chemical Society in London.

9.5   Crosland, Maurice P. *Historical Studies in the Language of Chemistry.* Cambridge, MA: Harvard University Press, 1962. 406 pp.

• Robert Boyle, the seventeenth century chemist whose name is familiar to all chemists today because of the important research that he did, recognized the significance of a precise nomenclature to the development of a science. Crosland's scholarly work traces the evolution of chemical nomenclature from the days of the classical philosophers and the alchemists (whose ideas about nomenclature were very different from Boyle's) to the important congresses of the nineteenth and early twentieth century which laid the foundations upon which contemporary nomenclature is based. An extensive bibliography is included.

9.6   *Enzyme Nomenclature.* Amsterdam: Elsevier, 1973. 443 pp.

• Subtitle: Recommendations (1972) of the Commission on Biochemical Nomenclature on the Nomenclature and Classification of Enzymes together with their Units and Symbols of Enzyme Kinetics.
The Commission is a body under the joint auspices of the International Union of Pure and Applied Chemistry and the International Union of Biochemistry. The nomenclature described in this book is based on that which has been in fairly general use in journals, textbooks, and other publications since 1961, but the situation before that has been described as chaotic in view of the multiplicity of names that were in use. Each enzyme in the list has an enzyme code number, a systematic name which depends upon the reaction that it catalyzes

and a non-systematic name which is referred to in this book as the recommended name. Other names are also sometimes listed. The enzymes are listed in numerical.order, and there is an index by systematic, recommended, and other names.

9.7   International Union of Pure and Applied Chemistry. Commission on Macromolecular Nomenclature. Pure Appl. Chem., 48: 373-85 (1976). "Nomenclature of Regular Single-Strand Organic Polymers."

• The IUPAC recommendations in this paper apply to organic polymers that can be described as a chain of regularly repeating structural units.

9.8   International Union of Pure and Applied Chemistry. Commission on symbols, Terminology, and Units. *Manual of Symbols and Terminology for Physicochemical Quantities and Units.* London: Butterworths, 1970.

• Included are names and symbols for mechanical, molecular, thermodynamic, chemical kinetic, electrochemical, optical, and other quantities, as well as mathematical and spectroscopic symbols, symbols for elements, nuclides, and particles, and sign conventions in electrochemistry.

9.9   International Union of Pure and Applied Chemistry. Commission on the Nomenclature of Inorganic Chemistry. Pure Appl. Chem., 30: 683-710 (1972). "Nomenclature of Inorganic Boron Compounds."

• Boron compounds exhibit structural features that are not common among the compounds of other elements. They also constitute one of the classes of compounds that is very extensively studied by inorganic chemists. This paper is more detailed than the rather short section on boron compounds in Ref. 9.10, and it includes some good illustrations which clarify the three-dimensional structural features.

9.10   International Union of Pure and Applied Chemistry. Commission on the Nomenclature of Inorganic Chemistry. *Nomenclature of Inorganic Chemistry.* 2d ed. London: Butterworths, 1971. 110 pp.

• The first edition (the 1957 Rules) was published in 1959. The second edition (the 1970 Rules) is a revision and extension, and is now the definitive guide to inorganic nomenclature, including elements, compounds in general, ions and radicals, iso- and hetero-polyanions, acids, salts and salt-like compounds, coordination compounds, addition compounds, crystalline phases of variable composition and polymorphism.

9.10.1 International Union of Pure and Applied Chemistry. Commission on the Nomenclature of Inorganic Chemistry. *How to Name an Inorganic Substance.* Oxford. Pergamon, 1977. 36 pp.

• The Commission prepared this text to be used in conjunction with *Nomenclature of Inorganic Chemistry* (Ref. 9.10) and other authoritative inorganic nomenclature sources. There is a table based on formulas, and corresponding to each formula (as applicable), the name is given for: a) the uncharged atom, molecule or radical; b) the cation or cationic radical; c) the anion; d) the ligand; and e) the prefix in substitutive nomenclature. Authoritative sources are cited in each case. There are six published sources serving as authorities, including an appropriate volume for the eighth collective period of *Chemical Abstracts* and one for the ninth.

9.11 International Union of Pure and Applied Chemistry. Commission on the Nomenclature of Organic Chemistry. Pure Appl. Chem., *45*: 11-30 (1976). "Rules for the Nomenclature of Organic Chemistry. Section E: Stereochemistry."

• A footnote on page 13 points out that "these rules may be called the IUPAC 1974 Recommendations for Section E, Fundamental Seterochemistry." The Commission also points out that rules already exist for specialized classes of compounds such as peptides and carbohydrates.

The Commission recognizes that some of the rules they present are still subject to controversy. The lack of standardization in this nomenclature has long been a source of frustration to the users of chemical literature, and even in recently published reference works, such as the *Atlas of Stereochemistry* by Klyne and Buckingham (Ref. 13.15.5) non-systematic, and even inconsistent, nomenclature may be encountered.

Stereochemistry is defined as chemistry in three-dimensional space. All molecules are three-dimensional, but stereochemical nomenclature becomes important only when it is necessary to distinguish one molecule from others on the basis of the arrangement of atoms in three dimensions. This spacial aspect of molecular structure has long been recognized, but it is particularly significant for biologically active molecules. Many chemists have chosen to ignore it or attach little importance to it until recently. Consequently, there are many compounds which have more than one *Chemical Abstracts* registry number—one for each of the distinct stereochemical forms, and one for the compound when it was reported without any indication of its stereochemistry. This is an important factor for the user of the CASIA database to bear in mind (Ref. 5.4).

A bibliography of journal articles on stereochemical nomenclature is given on pages 151-152 of *Stereochemistry. An Introductory Programmed Text*, published by Burgess Publishing Company, Minneapolis, 1976.

9.12   International Union of Pure and Applied Chemistry. *Nomenclature of Organic Chemistry*. 3rd ed. London: Butterworths, 1971. 337 pp.

- Section A: Hydrocarbons.

     Section B: Fundamental heterocyclic systems.

     Section C: Characteristic groups containing carbon, hydrogen, oxygen, nitrogen, halogen, sulfur, selenium and/or tellurium.

The definitive rules for naming organic compounds which fit within the parameters of the three sections are given in this book. If the organic compound is a ligand in a coordination compound, these rules must be used in conjunction with the IUPAC nomenclature for inorganic compounds (see Ref. 9.10). Certain specialized classes of organic compounds, such as steroids, have had rules formulated for them independent of the rules in this book, and in general they should be named according to their own specialized nomenclature. It is the intention of this guide to include mention of the latest published definitive or tentative nomenclature rules for major classes of compounds, but not for all of the many classes of less general significance. For sections D and F, see Ref. 9.14 Nos. 31 and 53, and for section E, see Ref. 9.11.

9.13   International Union of Pure and Applied Chemistry and International Union of Biochemistry. IUPAC Commission on the

Nomenclature of Organic Chemistry and IUPAC-IUB Commission on Biochemical Nomenclature. Pure Appl. Chem., *31*: 285-322 (1972). "Definitive Rules for Nomenclature of Steroids."

• A reprint of this paper was published in 1972 by Butterworths of London.

9.14 International Union of Pure and Applied Chemistry. *Information Bulletin. Appendices on Provisional Nomenclature, Symbols, Units and Standards.* Oxford: International Union of Pure and Applied Chemistry, 1970-. Irregular.

• The titles of a few of the issues are listed below, attention having been given to those which bear on particularly important aspects of nomenclature and which have not been superseded by definitive rules. Some of these issues have also appeared in other publications.

No. 19. Rules for the Nomenclature of Carotenoids. (1972)

No. 23. Symbols for Amino-Acid Derivatives and Peptides. (1972)

No. 31. Nomenclature of Organic Chemistry: Section D. (1973) Provisional rules covering organic derivatives of elements other than those named in Section C. (See Ref. 9.12.)

No. 45. List of Trivial Names and Synonyms (for substances used in analytical chemistry). (1975)

No. 46. Nomenclature of α-Amino Acids. (1975)

No. 53. Nomenclature of Organic Chemistry: Section F—Natural Products and Related Compounds. (1976)

No. 55. Recommendations for the Naming of Elements of Atomic Numbers Greater than 105. (1976)

9.15 Royal Society of London. Symbols Committee. *Quantities, Units, and Symbols.* 2d ed. London: The Royal Society, 1975. 54 pp.

• This authoritative guide to the use of units and symbols in the physical sciences has been distilled from the publications of recognized bodies such as the International Organization for Standardization and the International Union of Pure and Applied Chemistry.

9.16   *Unified Numbering System for Metals and Alloys.* Warrendale,
       PA: Society of Automotive Engineers, 1975.

   • Different standards organizations, societies, manufacturers,
   etc. have individual coding systems for metals and alloys. The
   unified numbering system (UNS), a joint activity of the Society
   of Automotive Engineers and the American Society for Testing
   and Materials, is not an attempt to impose a single system to
   supersede the others, but rather provides a means for associat-
   ing the numbers for the same material from different systems.
      Each UNS number consists of a letter followed by five digits.
   The letters characterize the major metal families, such as A for
   aluminum and its alloys, and T for tool steels. The book is
   arranged by UNS number. Each entry gives a description of the
   metal, its chemical composition, and a listing of the codes that
   have been applied to it by different organizations. Twelve
   organizations, including some departments of the U.S. Federal
   Government, are represented. The lists of codes of these organizations are presented at the end of the book, and as each
   code has the UNS number associated with it, these lists constitute an index to the UNS number entries.

9.17   Young, John A. J. Chem. Doc., *14:* 98-100 (1974) "Revised No-
       menclature for Highly Fluorinated Organic Compounds."

   • Highly fluorinated organic compounds have become important recently in the science of chemistry and its applications. The highly useful polymer with the tradename Teflon is
   a well-known example. The traditional systematic organic
   nomenclature which requires that all non-hydrogen substituents on the skeleton of the molecule be specifically named is
   very awkward for this class of compounds. Young points out
   that the less awkward nomenclature that he describes has been
   authorized by the American Chemical Society.

10

Dictionaries
and Encyclopedias

For the field of chemistry, there are a few dictionaries that are alphabetical listings which provide definitions of chemical terms. However, chemical dictionaries often also include biographical sketches, synonyms for trade names, and even physical data for specific chemical substances. In fact, there is no sharp distinction between chemical dictionaries and encyclopedias; many of the works in this section are a bit of both. Language dictionaries, as well as books which deal strictly with nomenclature, or which are primarily compilations of data, are listed elsewhere.

There are many specialized dictionaries and glossaries which may be of use to the chemist, and in libraries which make use of the Library of Congress subject headings, these may usually be identified under the specific subject which applies, with the subheading "dictionaries." This same subheading is also commonly used for encyclopedias.

10.1 *Concise Chemical and Technical Dictionary.* Edited by Harry Bennett. 3rd ed. New York: Chemical Publishing Company, 1974. 1175 pp.

- There are over 50,000 very brief entries in this dictionary, which is particularly valuable because of the large number of trade-named products that are identified. Many common compounds are characterized by their formulas and physical properties. The nomenclature is essentially that which was used in the 1930's. Technical terms such as "abrasive" and "zygote" are very briefly defined, and the technological meanings of some common English words, such as "absolute" are also supplied. It must be emphasized that this dictionary is intended for use in connection with industrial and technological aspects of chemistry.

There are several auxiliary tables, and one of the most useful is a guide to the pronunciation of chemical words, a feature which is surprisingly rare in chemical dictionaries. *An Explaining and Pronouncing Dictionary of Scientific and Technical Words* (2d ed. London: Longmans, Green, 1953) is a good source for British pronunciation.

10.2   *The Condensed Chemical Dictionary.* 9th ed. Revised by Gessner G. Hawley. New York: Van Nostrand Reinhold, 1977. 957 pp.

• This standard reference work includes definitions of terms, data on chemicals of technical importance, identification of trademarked products, abbreviations, and other information, all in a single alphabet. Emphasis is on entries of practical importance in chemical technology. For example, in many of the entries on specific chemicals, hazard information and shipping regulations are included.

10.3   *The Encyclopedia of Chemistry.* Edited by Clifford A. Hampel and Gessner G. Hawley. New York: Van Nostrand Reinhold, 1973. 1198 pp.

• The articles are of intermediate length, on fairly broad topics (e.g., classes of compounds rather than individual compounds).

10.4   Flood, W.E. *The Origins of Chemical Names.* London: Oldbourne, 1963. 238 pp.

• Part I: The chemical elements. Part II: Chemical compounds, minerals, and other substances of chemical interest. Each entry gives not only the etymology of the name, but also a brief history of the discovery or initial preparation of the substances. Some old names that are no longer used are also included.

10.5   *Hackh's Chemical Dictionary.* 4th ed. Edited by Julius Grant. New York: McGraw-Hill, 1969. 738 pp.

• Most of the nearly 55,000 very concise entries are for specific substances, but some concepts, name reactions, etc., are included. This dictionary presents both British and U.S. usage.

While the emphasis is on pure and applied chemistry, there are numerous entries from related sciences, including physics, mineralogy, pharmacy, and others.

10.6 Hampel, Clifford A., and Gessner G. Hawley. *Glossary of Chemical Terms.* New York: Van Nostrand Reinhold, 1976. 281 pp.

 • This is a particularly valuable reference work for the undergraduate in chemistry. The brief entries, which usually include more than a definition, are of commonly encountered terms from all fields of chemistry.

10.7 *International Encyclopedia of Chemical Science.* Princeton, NJ: Van Nostrand, 1964. 1331 pp.

 • Although this book is referred to as an encyclopedia, the entries are quite brief. However, the range of topics is very broad, with emphasis on topics of interest in academic programs. An important feature is the inclusion of numerous analytical testing methods. Indexes are in French, German, Spanish, and Russian.

10.8 *Kingzett's Chemical Encyclopaedia. A Digest of Chemistry and Its Industrial Applications.* 9th ed. D.H. Hey, general editor. Princeton, NJ: Van Nostrand, 1966. 1092 pp.

 • Entries, which tend to be only one or two paragraphs long for the most part, define individual substances as well as classes of substances such as "alum" or "anti-freeze agents." Trivial and trade names are included. There are also a few long articles, such as the one on nuclear chemistry and radioactivity which occupies sixteen pages. The longer articles include references to textbooks and other secondary sources. In short, this book combines the characteristics of an elementary encyclopedia of general chemistry and a dictionary of common chemical substances in a single alphabet. It is not for the specialist.

10.9 *A New Dictionary of Chemistry.* 4th ed. Edited by L. Mackenzie Miall and D.W.A. Sharp. New York: Wiley, 1968. 638 pp.

- This dictionary is intended for the non-specialist. It is one of the best chemistry dictionaries that an academic library can have for basic definitions, including, but not limited to, terms encountered in the undergraduate organic, inorganic, and physical chemistry courses. The entries are brief, accurate, and clear, and accompanied by molecular structural diagrams when appropriate.

10.10 *Thorpe's Dictionary of Applied Chemistry.* 4th ed. London: Longmans, 1937-1956. 12 vol.

- Thorpe's has been a standard encyclopedic dictionary of chemistry since the first three-volume edition (1890-1893). For most purposes, it is not the first dictionary to look into for definitions or other information on a contemporary subject. However, the growth of chemical information has been of such a magnitude that many topics that were in Thorpe's can no longer be carried by modern dictionaries. The articles are scholarly, and many of them are lengthy and well referenced. "Applied chemistry" in the case of this dictionary includes academic laboratory chemistry in the fields of analytical, organic and inorganic chemistry. When there is reason to believe that a subject was a topic under investigation by applied chemists during the earlier part of this century, the appropriate edition of Thorpe's might be a good place to start a literature search, both for an overview of early thoughts and for references to the early literature.

ANALYTICAL CHEMISTRY

10.11 *Encyclopedia of Industrial Chemical Analysis.* Edited by Foster Dee Snell and Clifford L. Hilton. New York: Interscience, 1966-1974. 20 vol.

- The first three volumes deal with general techniques. Volumes 4-19 treat the analyses of specific substances or classes of substances, from ablative materials to zinc. Within these later volumes, there are a few articles on specific methods of analysis. Volume 20 is an index to Volumes 4-19. The articles are long and detailed, accompanied by tables, illustrations, bibliographies, and most important, precise instructions in methods of analysis. This is a major reference work which can be used to obtain background information about the composition and analysis of materials of commercial importance, such as coffee, silicon compounds, or soap. It can also be used as a practical manual in the industrial analytical laboratory.

CHEMICAL ENGINEERING AND TECHNOLOGY

10.12 *Chemical and Process Technology Encyclopedia.* Editor-in-Chief: Douglas M. Considine. New York: McGraw-Hill, 1974. 1261 pp.

• Articles, written by experts, vary in length, but tend to be relatively short, and are intended primarily for the scientist, engineer or technical manager. However, they are clear and well-written, and can for the most part be understood by anyone who has had a couple of years of college chemistry. Both traditional and newly significant aspects of the subjects are included. Some articles have bibliographies, a very extensive "book" bibliography being provided for the article on "chemical engineering." A classified index, plus a very extensive alphabetical subject index, enhance the utility of the encyclopedia.

10.13 *Chemical Technology: An Encyclopedic Treatment.* New York: Barnes and Noble, 1968-75. 8 vol.

• Lengthy articles, some of which can be understood with only a minimal chemical background, deal with materials in everyday use. The volume titles indicate the nature of this work.
Vol. 1. Air, water, inorganic chemicals, and nucleonics. Vol. 2. Non-metallic ores, silicate industries and solid mineral fuels. Vol. 3. Metal ores and metals. Vol. 4. Petroleum, organic chemicals and plastics. Vol. 5. Natural organic materials and related synthetic products. Vol. 6. Wood, paper, textiles and photographic materials. Vol. 7. Vegetable food products and luxuries. Vol. 8. Edible oils and fats and animal food products. This volume includes an index to the entire set.

10.14 *Encyclopedia of Chemical Processing and Design.* Edited by John J. McKetta and William A. Cunningham. New York: Dekker, 1976-. Multivolume.

- The emphasis is on the practical design of equipment, systems and controls, and on plant construction, in the chemical industry. The majority of articles are each devoted to a single important industrial compound or class of related compounds, though there are a few entries on general processes. This is a fairly advanced encyclopedia, and will be of use mainly to the professional or student in chemical engineering.

10.15 Gardner, William, and Edward I. Cooke. *Chemical Synonyms and Trade Names. A Dictionary and Commercial Handbook.* 7th ed. London: The Technical Press, 1971. 689 pp.

- This is a standard source, international in coverage, which identifies chemical substances in commerce and technology, primarily by indicating their uses and chemical composition. In many cases no systematic chemical name is given in the entry.

10.16 Heinisch, K.F. *Dictionary of Rubber.* New York: Wiley, 1974. 545 pp.

- Included are a variety of terms connected with rubber, from sap to finished product, as well as instrumentation, names of people, chemicals used in rubber processing, etc. Some trade names are included, and they are coded to a directory of producers and marketing organizations which is appended at the end. References to ASTM or other standards are provided in a few entries.

10.17 *Kirk-Othmer Encyclopedia of Chemical Technology.* 2d ed. New York: Wiley, 1963-70. 22 vol.

- Approximately half the articles deal with chemical substances or classes of substances. Others are on fundamental principles, unit operations, methods of analysis, and similar topics. The articles are long, authored by experts, and are in general at an advanced level, suitable for the professional chemical engineer. Detailed tables and graphs, related to the technical, or on occasion, economic aspects of the topics enhance the utility of the individual articles as excellent ready-reference sources. The authoritative bibliographies that accompany each article are particularly noteworthy for their references to the patent literature.
A supplement volume was issued in 1971, and in the following year an index to the entire set, including the supplement,

was published. The first volume of the completely revised third edition appeared in 1978.

10.18 Simons, Eric N. *A Dictionary of Alloys*. London: Muller, 1969. 191 pp.

- Most of the entries are brief, giving composition and uses. Most steels and alloy cast irons have been omitted, but otherwise extensive coverage is given to British and American alloys. Proprietary names are included, and the author also provides definitions for alloy names that correspond to ranges of composition, such as "electrum" and "typewriter alloys."

10.19 Sittig, Marshall. *Inorganic Chemical and Metallurgical Process Encyclopedia*. Park Ridge, NJ: Noyes Development Corp., 1968. 883 pp.

- This volume, a companion to the author's *Organic Chemical Process Encyclopedia* (Ref. 10.20), presents flow diagrams and descriptions of the industrial preparation of nearly 2,000 inorganic chemicals. Most of the information is based on U.S. patents, to which references are given.

10.20 Sittig, Marshall. *Organic Chemical Process Encyclopedia*. 2d ed. Park Ridge, NJ: Noyes Development Corp., 1969. 712 pp.

- This unique encyclopedia relies on graphics to provide overviews of the production processes for industrial chemicals, mainly petrochemicals. One page is allotted to each entry, and nearly half of that page is devoted to a detailed flow diagram for the process. The chemical equation, feed materials, coproducts, catalyst (if any), reactor type, temperature, and similar data are provided, as well as a reference to a U.S. patent. This volume is a companion to Sittig's *Inorganic Chemical and Metallurgical Process Encyclopedia* (Ref. 10.19).

10.21 *Ullmans Encyklopädie der Technischen Chemie*. 4th ed. Weinheim/Bergstr.: Verlag Chemie, 1972-. Multivolume.

- The first volume (general principles) includes topics such as thermodynamics, fluid dynamics, and the optimization of chemical reactions. General chemical engineering operations and apparatus are the subjects of the second and third volumes.

Most of the fourth volume deals with the application of the computer to chemical plant operations. There is also a chapter on documentation. The alphabetical sequence of entries is to begin with volume 6. The articles are entirely in German, and are written for the professional chemical engineer. However, each volume includes a detailed index in English. This is perhaps the most important modern encyclopedia of chemical engineering, and it is particularly noteworthy for the excellent illustrations and flow diagrams of chemical operations. Bibliographies of patents, research papers, and other literature accompany the articles.

ELECTROCHEMISTRY

10.22 Davies, C.W. and A.M. James. *A Dictionary of Electrochemistry*. London: Macmillan, 1976. 246 pp.

• The alphabetically arranged entries are short essays, not simply definitions, of electrochemical terms and topics. However, the subjects are not integrated into large, topical essays, as they are in Hampel's *Encyclopedia of Electrochemistry* (Ref. 10.24). This dictionary will be especially useful for the chemist who is not an electrochemist, but who needs a quick overview of an electrochemical concept, process, or device.

10.23 *Encyclopedia of Electrochemistry of the Elements*. edited by Allen J. Bard. New York: Dekker, 1973-. Multivolume.

• Different experts contributed the individual chapters, each of which is devoted to a single element or group of closely related elements. The arrangement is inspired by that in Gmelin (see Chapter 14) but is not identical to it. The user is advised to look at the beginning of one of the most recently published volumes and consult the organizational chart to find which volume and chapter cover a particular element.
   The chapters are generally organized in the same way: introduction and standard potentials, voltammetric characteristics, kinetic parameters and double layer properties, survey and critical review of the electrochemical studies and applications. Tables of standard electrode potentials, polarographic data, etc. are given, and each chapter concludes with an extensive bibliography.

10.24 Hampel, Clifford A. *The Encyclopedia of Electro-Chemistry.* New York: Reinhold, 1964. 1206 pp.

• Fairly long articles, including biographies, have been contributed by specialists. Most articles include a list of literature references.

INORGANIC CHEMISTRY

10.25 *The Encyclopedia of the Chemical Elements.* Edited by Clifford A. Hampel. New York: Reinhold, 1968. 849 pp.

• A relatively long article has been prepared for each element, outlining the history, occurrence, preparation, chemical, physical, and hygienic properties, and applications. A few key references to primary and secondary publications accompany each article.

NAMED REACTIONS, LAWS, ETC.

10.26 Ballentyne, D W.G., and D.R. Lovett. *A Dictionary of Named Effects and Laws in Chemistry, Physics, and Mathematics.* 3rd ed. London: Chapman and Hall, 1970. 335 pp.

• The brief entries, alphabetically arranged, provide the minimum information needed to identify the chemical reactions, reagents, synthetic and analytic procedures, theories, laws, physical constants, etc., that are named after individual scientists. Most of these are "textbook" expressions, and they appear often in the literature, commonly without explanation or bibliographic citation. They are, in fact, a part of the vocabulary of the physical sciences.

Named units are treated in a separate alphabetical listing of entries in an appendix at the end. There appears to be little advantage to listing them separately, but overall this extremely useful dictionary is commendably easy to use and to the point.

10.27 Denney, Ronald C. *Named Organic Reactions.* London: Butterworths, 1969. 252 pp.

• Each of the seventy-two chapters is a short, but encyclopedic treatment of an important name reaction, such as the Claisen Rearrangement or the Friedel-Crafts Reaction. The nature of each reaction is illustrated, followed by brief discussions of the historical development, mechanism, general reaction conditions, modifications, and applications. Each chapter ends with a list of references. Although the amount of detail is insufficient for guidance in laboratory procedures, the articles in *Named Organic Reactions* are very useful for quick overviews of the reactions.

10.28 Gowan, J.E. and T.S. Wheeler, *Name Index of Organic Reactions*. New York: Interscience, 1960. 293 pp.

   • There are 739 named reactions presented here, far more than in Denney's *Named Organic Reactions* (Ref. 10.27). However, the entries in this book are much briefer—usually just sufficient to define each reaction under consideration. Many of the named reactions in this book are obscure. There are a lot of reactions which have two names associated with them, such as Friedel-Crafts, and the authors have been kind enough to provide cross references from the second names in such cases. In addition, there are reaction type and general indexes.

10.29 Krauch, Helmut, and Werner Kunz. *Organic Name Reactions*. New York: Wiley, 1964. 620 pp.

   • Concise information is provided for approximately five hundred reactions. The user should be aware of additional references provided in an addendum near the end of the book.

ORGANIC CHEMISTRY

10.30 *Dictionary of Organic Compounds*. 4th ed. New York: Oxford University Press, 1965-. 5 vols. plus annual supplements.

   • This is one of the most extensive compendia of basic data and information on organic compounds in the English language, and it is noteworthy for its coverage of natural products as well as laboratory reagents and important compounds that have been synthesized in the lab. Although this dictionary does not cover nearly as many compounds as Beilstein's Handbuch (see Chapter 14), it is generally more up to date.

   Basic data, such as molecular weight, formula, melting or boiling point, and refractive index are given for each compound, and the entries usually also include structural formulas. Each entry is accompanied by important literature references, usually to methods of preparation.

The nomenclature very often does not correspond to that used by *Chemical Abstracts*. The inversion of names that is common in many indexes, including those of *Chemical Abstracts*, is not practiced in this dictionary. Certain types of compounds do not receive separate entries. For example, esters and amides are usually listed under their parent acids. Numerous cross references have been provided, and there is a formula index to the initial five-volume set. Annual supplements follow the same format, and these are cumulated into five-year supplements. Formula indexes accompany the supplements.

10.31 *The Merck Index. An Encyclopedia of Chemicals and Drugs.* 9th ed. Rahway, NJ: Merck and Co., 1976. 1313 pp., plus appendices and indexes.

• The Merck Index is highly esteemed and heavily used by research chemists for basic information on organic compounds, especially drugs and biologically active compounds. Nearly 10,000 such compounds are treated (nearly 1,000 more than in the previous edition) with structural diagrams, key references to the primary literature, and important properties, use, toxicity, and other information.

The names of the compounds, in general, correspond to common usage, but in most cases the uninverted form of the ninth collective *Chemical Abstracts* name is provided in bold-faced italics, within the entry, and other synonyms may also be given. Reference to an entry from synonyms is achieved *via* a lengthy name index, rather than through cross references within the body of the text. There is also a formula index. Among the more important of the appendices are a dictionary of name reactions, a table of radioisotopes, a company register, a table of saturated solutions, a table of isotonic solutions and a listing of *Chemical Abstracts* registry numbers for the compounds covered in the Merck Index.

10.32 Patai, Saul. *Glossary of Organic Chemistry, Including Physical Organic Chemistry.* New York: Interscience, 1962. 227 pp.

• The inclusion of physical organic chemistry distinguishes this dictionary from others written for the organic chemist. Name reactions are included among the entries, but individual compounds are not. An interesting feature is the provision of references to pages in major textbooks, monographs, and review periodicals where the specific topics are treated.

10.33 Synthetic Organic Chemical Manufacturers Association. *SOCMA Handbook. Commercial Organic Chemical Names.* Washington: American Chemical Society, 1966. 935 pp.

• Structural formulas and synonyms, including trade names, are given for over 6,000 pure organic compounds, mixtures, and polymers.

PHYSICAL CHEMISTRY

10.34 James, Arthur M. *A Dictionary of Thermodynamics.* London: Macmillan, 1976. 262 pp.

• In spite of the relatively advanced level of this dictionary/ encyclopedia, people with only a minimal college exposure to the physical sciences should be able to comprehend most of the basic thermodynamic topics treated. The entries, many of which extend over several pages, are accompanied by mathematical and chemical equations, tables, and graphs when appropriate.

POLYMER CHEMISTRY

10.35 *Encyclopedia of Polymer Science and Technology.* New York: Wiley, 1964-72. 16 vol.

• The lengthy, authoritative articles were prepared by specialists and discuss specific natural and man-made polymeric materials, polymer properties, engineering and laboratory processes, analytical methods and uses. The article on "Literature of Polymers" is an extensive bibliographic guide. Volume 16 is an index to the set. The first supplement, published in 1976, includes some revisions of earlier articles, and numerous new subjects.

10.36 International Union of Pure and Applied Chemistry. Commission on Macromolecular Nomenclature. Pure Appl. Chem., 40:479-91 (1974). "Basic Definitions of Terms Relating to Polymers."

• Terms related to polymer structures and the polymerization process are defined.

SPECTROSCOPY

10.37 Denney, Ronald C. *A Dictionary of Spectroscopy*. New York: Wiley, 1973. 161 pp.

• This dictionary, intended for the general spectroscopist and the undergraduate, deals with many forms of spectroscopy of interest to the chemist, including mass, infrared, ultraviolet, nuclear magnetic resonance, Raman and fluoresence spectroscopy. The brief entries often include excellent illustrations and citations to the primary literature. Apparatus, techniques, theory, and applications supply the terms to be defined. Users will be pleased with the clear, concise explanations of how various spectroscopic properties of substances are determined, usually with an indication of the units in which the properties are reported. Terminology is in accordance with the recommendations of the Symbols Committee of the Royal Society of London and the International Union of Pure and Applied Chemistry, and British spelling is used.

11
Language
Dictionaries

Only a few languages are really important in contemporary chemistry. In terms of volume of publication, English is first and Russian second. A significant but relatively small volume of publication occurs in French and German, but these languages are particularly important because of the vast amount of earlier chemistry, including most of the foundations of modern organic chemistry, published in them. Some of the reference works that are indispensable for chemical research are in German, as testified by several entries in this guide.

Italian, Polish and a few other European languages cannot be completely ignored because of the occasional significant research paper, and because of trade and similar publications that may be important to commercial organizations. However, Japanese, in recent years, has overtaken them to become one of the five most important languages in chemistry.

Most of the more important Russian language research journals, and many of those in Japanese, are now translated, cover to cover, into English. However, there is always a time lag between the appearance of the original and the translation.

In this guide, emphasis has been placed on those dictionaries that may aid in translating from foreign languages into English. For many languages, there are no language dictionaries specifically for chemistry but they do exist for the most important ones. To locate dictionaries for translating foreign languages into English in the library, the user must be aware of the subject-heading conventions employed by the library. Most academic libraries in the U.S. use Library of Congress subject headings. As an example of the way Library of Congress subject headings apply to language dictionaries, a Turkish-English dictionary receives the subject entry:

Turkish language—Dictionaries—English.

However, it should be noted that language dictionaries that include more than two languages (the so-called polyglot dictionaries) have the entry:

Dictionaries, polyglot

if they are general, or entries like:

Chemistry—Dictionaries—Polyglot

if they are in a specific subject area.

In the index to this guide, entries are provided for *all* of the languages of a dictionary (except English) in the form "French language," "Chinese language," etc.

11.1   Aghina, Luisa, *Dizionario Tecnico Italiano-Inglese, con Particolare Riferimento alla Industria Chimica*. Florence: Vallechi, 1961. 431 pp.

   • Italian-English.

11.2   Callaham, Ludmilla Ignatiev. *Russian-English Chemical and Polytechnical Dictionary*. 3rd ed. New York: Wiley, 1975. 852 pp.

   • This is the most important Russian-English chemical dictionary. Emphasis is on scientific rather than technological terms, but there are many general, non-chemical terms as well.

11.3   Carpovich, Eugene A. and Vera V. Carpovich. *Russian-English Chemical Dictionary: Chemistry, Physical Chemistry, Chemical Engineering, Materials, Minerals, Fuels, Petroleum, Food Industry, Pharmacology*. 2d ed. New York: Technical Dictionaries, 1963. 352 pp.

   • Emphasis is on technology. Chemical formulas are included for some common compounds.

11.4   Chu, Chi-hsüan. *Ying Yung Hua Hsüeh Tzú Tien*. Hong Kong: Commercial Press, 1954. 1111 pp.

   • This is a Chinese-English and English-Chinese dictionary, mainly of names of chemical compounds. Reprinted in 1964.

11.5 DeVries, Louis and Helga Kolg. *Wörterbuch der Chemie und der Chemischen Verfahrenstechnik.* Weinheim/Bergstr.: Verlag Chemie, 1970-1972. 2 vol.

- Vol. 1 German-English; Vol. 2 English-German. Fields such as biochemistry, biophysics, physiology, atomic physics and electronics are included, as well as core chemistry.

11.6 *Dictionary of Chemistry and Chemical Technology in Six Languages: English, German, Spanish, French, Polish, Russian.* Edited by Z. Sobecka, W. Choínsky, and P. Majorek. Oxford: Pergamon, 1966. 1325 pp.

11.7 *Elsevier's Dictionary of Industrial Chemistry in Six Languages: English/American, French, Spanish, Italian, Dutch, and German.* Compiled by A.F. Dorian. Amsterdam: Elsevier, 1964. 2 vol.

- Arrangement is alphabetical, in English, with each entry including an English language definition, and the five foreign language equivalents of the term. There are indexes in each of the five foreign languages.

11.8 Ernst, Richard, and Ingeborg Ernst von Morgenstern. *Dictionary of Chemistry, Including Chemical Engineering and Fundamentals of Allied Sciences.* Wiesbaden: Brandstetter, 1961-63. 2 vol.

- Vol. 1, German-English; Vol. 2, English-German.

11.9 Fouchier, Jean, and F. Billet. *Chemical Dictionary.* 2d ed. Amsterdam: Netherlands University Press, 1961. 1271 pp.

- English, French, and German.

11.10 Hashimoto, Yoshirō. *New English-Japanese and Japanese-English Dictionary for Chemist.* Tokyo: Maruzen, 1970. 528 pp.

11.11 Hoseh, Mordecai, and Melanie L. Hoseh. *Russian-English Dictionary of Chemistry and Chemical Technology.* New York: Reinhold, 1964. 522 pp.

- This excellent dictionary confines itself to the chemical meanings of terms.

11.12 Japan. Ministry of Education. *Japanese Scientific Terms, Chemistry*. Revised and enlarged edition. Tokyo: Chemical Society of Japan, 1974. 633 pp.

11.13 Kagaku Kōgaku Kyōkai. *Kagaku Kōgaku Jiten*. Tokyo: Maruzen, 1974. 551 pp.

• Because English terms are included with each entry, and there is an English language index, this Japanese dictionary of chemical engineering can also function as a Japanese-English and English-Japanese glossary.

11.14 Neville, Hugh Henry, N.C. Johnston, and G.V. Boyd. *A New German/English Dictionary for Chemists*. Princeton, NJ: Van Nostrand, 1964. 330 pp.

• This dictionary includes many non-technical terms, as well as basic chemical terms, and a few words from related sciences. It is therefore a particularly practical aid for the chemist who needs to read articles or other documents in that language, and who does not know German very well.

11.15 Patterson, Austin McDowell. *A French-English Dictionary for Chemists*. 2d ed. New York: Wiley, 1954. 476 pp.

• This French-English chemistry dictionary is very extensive in coverage, and has been a standard reference work for many years, as has the German-English dictionary by the same author in the entry which follows.

11.16 Patterson, Austin McDowell. *A German-English Dictionary for Chemists*. 3rd ed. New York: Wiley, 1950. 541 pp.

11.17 Reid, Ebenezer Emmet. *Chemistry Through the Language Barrier; How to Scan Chemical Articles in Foreign Languages with Emphasis on Russian and Japanese*. Baltimore: Johns Hopkins Press, 1970. 138 pp.

• Appendices include lists of common elements and other terms with their English equivalents in French, Spanish, Italian, Romanian, German, Dutch, Swedish, Danish, Hungarian, Finnish, Polish, Czech, Russian, and Japanese.

11.18 *Russian, Chinese, English Chemical and Technical Dictionary.* London: Scientific Information Consultants, 1965. 279 pp.

- This version of the dictionary was compiled by specialists at the East China School of Chemical Technology. As it is arranged alphabetically by Russian terms, with no indexes, it is primarily of use to Westerners as a Russian-English dictionary, mainly for chemical compounds and common chemical terminology.

11.19 *Technical Dictionary of Spectroscopy and Spectral Analysis: English, German, French, Russian, with a Supplement in Spanish.* Edited by Heinrich Moritz and Tibor Torök. Oxford: Pergamon, 1971. 188 pp.

11.20 Wohlauer, Gabriele E.M. and H.D. Gholston. *German Chemical Abbreviations.* New York: Special Libraries Association, 1965. 63 pp.

- The German and English equivalents for each German abbreviation are given.

12
General Compilations of Data

Tabulations of data based on the physical measurements that are reported in journal articles and other primary publications can often save the chemist many hours of literature searching. On occasion, tabulations include data that are not to be found elsewhere in the literature. It should be noted that some compilations are comprised in whole or in part of tables that are taken from other secondary sources. These last are usually intended for individual purchases, and their value is primarily in the convenience to their owners.

There are many conflicting values of data in the literature. Only a few data compilations claim detailed authoritative evaluation of the data in them. The critical evaluation of data is an expensive task, and it must be done by people who are experts in the specific subject areas of the data. Government laboratories and standards organizations have undertaken or sponsored the critical evaluation of chemical data, and the publication of critically evaluated data compilations.

12.1 Aylward, G.H., and T.J.V. Findlay. *SI Chemical Data*. Sydney: John Wiley and Sons Australasia Pty. Ltd., 1971. 127 pp.

• All of the constants and property data in this book are in SI (Système International d'Unités) units, a coherent system of units that have been adopted by many countries. The basic SI units, and common derived SI units, are listed on page 1. For multiple and fractional decimal units, the system makes use of prefixes which are (usually) familiar from the metric system, to which the SI system is related. Conversion factors to other units commonly encountered are on pages 3 and 4.

This book is written for university students, and is limited to the basic units and common substances usually encountered in undergraduate chemistry courses. Most of the data

81

can be located in the large standard handbooks, though not necessarily in SI units. The small size and ease of arrangement make this a particularly handy reference work for thermochemical data of common organic and inorganic compounds, and for basic atom and bond-related data, such as average bond enthalpies, bond lengths, electronegativities, and ionization potentials.

12.2Basic Tables in Chemistry. Edited by Roy A. Keller. New York: McGraw-Hill, 1967. 400 pp.

• This book is intended primarily for the undergraduate, and most of the tables are reproduced from other sources, mainly the tenth edition of Lange's Handbook of Chemistry (Ref. 12.12). A few of the tables such as the color-code conventions and schematic symbols for electronic circuit components, are not commonly found in chemistry handbooks.

12.3Chemical Engineers' Handbook. 5th ed. Edited by Robert H. Perry and Cecil H. Chilton. New York: McGraw-Hill, 1973. 1954 pp.

• This standard reference work in chemical engineering is often the best first place to look for data on the physical properties of industrially important chemicals.

12.4CRC Handbook of Chemistry and Physics. 58th ed. Edited by Robert C. Weast. Cleveland: CRC Press, 1977. 2348 pp.

• The CRC Handbook is undoubtedly the best known and most heavily used chemistry handbook in the English-speaking world. A new edition appears annually, but usually with only minimal changes and additions to the immediately preceding edition. However, the editor notes in his preface that the 58th edition is more than 50 times the size of the first edition which was published in 1913. Since this is the first source that the chemist or student will often turn to for very basic data, it deserves careful analysis here. Extensive tables in mathematics and physics are provided, but attention here will focus on the chemical information.

There are three long tables of the physical properties of compounds; inorganic, in Section B, and organic and organometallic in C. Inverted names are usually used for the organic compounds, and the nomenclature tends to reflect common usage, with cross references in some cases from the systematic names. In all three tables, molecular weight, color, crystalline form, melting and boiling points, density, refractive index and solubility in water and other common solvents are provided, the solubility usually being simply a qualitative indication of whether or not the substance dissolves. No references are provided with the inorganic or organometallic tables. Beilstein, and on occasion other, references are included in the table of organic compounds, but there is no indication that this is the source of the data. The Handbook thus provides easy entry into Beilstein for many relatively common organic compounds. The organic table has indexes by increasing boiling and melting points. There are many tables of data useful in connection with traditional laboratory techniques, including gravimetric factors and their logarithms, solubility products, azeotropic data, preparation of laboratory reagents and decinormal solutions, calibration of volumetric glassware, and concentration-related properties of aqueous solutions of common substances.

There are also extensive vapor pressure tables for water and many other substances. The lengthy compilation of conversion factors is admirably easy to use.

Thermodynamics, dissociation constants, electrochemical data, isotopic information and much else found in this book is also available in numerous other handbooks, as well as specialized reference books. In many cases, the user should be aware that some specialized works will have more complete and accurate data than the general handbooks. The user is also advised to consult the *CRC Composite Index* (2d ed., 1977) for detailed subject indexing to this and other CRC handbooks.

12.5   Gordon, Arnold J., and Richard A. Ford. *The Chemist's Companion. A Handbook of Practical Data, Techniques, and References.* New York: Wiley 1972. 537 pp.

• In 1965, members of the American Chemical Scociety were asked "What physical, chemical, and mechanical properties of substances and systems do you seek most often in the literature?" According to the Preface, the *Chemist's Companion* covers most of the properties mentioned. The book is divided into the following broad sections: properties of molecular systems, properties of atoms and bonds, kinetics and energetics, spectroscopy, photochemistry, chromatography, experi-

mental techniques, mathematical and numerical information, and miscellaneous. Among the data not easily located in
other general handbooks are: bond lengths and interbond
angles; force constants; activation parameters; data for linear
free energy relationships, such as the Hammett Equation;
chemical shift and coupling constant correlation tables for
nmr; common solvents for crystallization; wavelength-wave-
number conversion; and character tables. Throughout this
handbook, there is a lot of practical information for the
laboratory or classroom, and when appropriate, suppliers are
mentioned. An example is the fairly extensive treatment of
atomic and molecular models.

12.6   Grasselli, Jeanette G., and William M. Ritchey. *Atlas of Spectral Data and Physical Constants for Organic Compounds*. 2d
ed. Cleveland: CRC Press, 1975. 6 vol.

   • This reference set is particularly valuable in that it provides a multitude of access points for the identification of an
organic compound, as well as entry to other tools of the
chemistry literature, including some major collections of spectra and Beilstein's *Handbuch*. Proton NMR, C-13 NMR, IR,
UV and mass spectral data, and physical constants are provided for about 21,000 compounds. The data in the *Atlas* have
been compiled from other secondary sources, and were not
critically evaluated. Hence, it is the indexing features and
convenience that insure this *Atlas* a prime place on the
reference shelves of chemistry libraries.
   There are nearly 200 pages of compound names in Volume
1 which serve as a name index to the data collections. Some
compounds are listed under both common and systematic
names. There are also tables that can aid in naming a compound, such as the common ring structures which are identified on pages 189-192.
   There is a permuted Wiswesser Line Notation index in
Volume 5. For those not familiar with the notation, there is a
dictionary of frequently found substructures. The *Atlas* may
help a chemist who has a spectrum or some physical data for a
compound he wishes to identify, through the indexes by the
various pieces of spectral and physical data.

12.7   *Handbook of Spectroscopy*. Edited by J.W. Robinson. Cleveland: CRC Press, 1974. 2 vol.

- Individual chapters, prepared by different authors, treat a range of topics that can broadly be considered as aspects of "spectroscopy." Most of the chapters include extensive tabulations of data, but a few are essentially review texts.

Nearly five hundred pages of Volume 1 are devoted to data from ESCA (Electron Spectroscopy for Chemical Analysis) and photoelectron spectroscopy. This constitutes one of the most extensive compilations for these two related fields. Another chapter is designed to aid the user in identifying molecular spectra that are encountered in flames. The first volume also contains extensive data related to x-ray spectroscopy.

The most noteworthy compilation in Volume 2 is the nmr chapter, which is modelled in format on Brügel's tables (Ref. 13.20.4).

Each volume is paged separately and has its own author and subject index. All authors who were cited and all compounds for which data are given are included in the indexes.

12.8   *Journal of Physical and Chemical Reference Data.* Washington: Published by the American Chemical Society and the American Institute of Physics for the National Bureau of Standards, 1972-. Quarterly.

- This journal is an important medium for the publication of up-to-date data. Very high standards are maintained. Some issues include Data Compilation Abstracts which review other sources of data. A few of the articles in this journal have presented data of such general utility that they have been given separate entries in this guide.

12.9   Kaye, G.W.C., and T.H. Laby. *Tables of Physical and Chemical Constants and Some Mathematical Functions.* 13th ed. New York: Wiley, 1966. 249 pp.

- This relatively small book necessarily includes only the most basic data in easy-to-use tables. References to more comprehensive works are often given.

12.10   Landolt, Hans Heinrich. *Zahlenwerte und Funktionen aus Naturwissenschaften und Technik. Neue Serie.* Edited by K.H. Hellwege. Berlin: Springer, 1961-. Multivolume.

• This "new series" is not intended as a seventh edition of the Landolt-Börnstein tables (12.11), and does not have an arrangement based on the rigid classification in the earlier volumes. Whereas the sixth edition was almost entirely in German, the new series uses both English and German.

Nearly all of the data of physics and some of the data of other sciences will at times be important to the chemist. Hence, there are many valuable tables in this series in addition to those noted here. Among those volumes of particular significance in chemistry are the following: 2:1 esr data and magnetic properties of free radicals; 2:2 and 2:8 magnetic properties of transition metal compounds; 2:3 luminescence of organic compounds; 2:4 molecular constants from microwave spectroscopy; 2:6 molecular constants from microwave, esr, and molecular beam spectroscopy; 2:7 geometric parameters of free molecules; 3:5 structure data of organic crystals; 3:6 structure data of elements and intermetallic phases; 3:7 structure data of inorganic compound crystals; 4:1 liquid densities. References to the primary sources are provided for all data.

12.11   Landolt, Hans Heinrich. *Zahlenwerke und Funktionen aus Physik, Chemie, Astronomie, Geophysik und Technik.* 6th ed. Berlin: Springer, 1950-1976. 4 vol. in 27.

• Usually referred to as the Landolt-Börnstein Tables, this compilation of critically evaluated data has long been recognized as one of the most authoritative reference sources. Volume I, atomic and molecular physics, and II, properties related to the state of aggregation, are of most interest to chemists.

Some particularly notable features of the various parts are the following: 1:1 data related to atomic spectra; electron distributions in atoms and ions: 1:2 molecular symmetry; parameters of molecular vibration and rotation; 1:3 electric dipole moments; molecular polarizabilities; 1:4 crystal data; 2:1 densities and compressibilities in the gaseous and condensed phases; 2:2a vapor pressures; freezing and boiling point constants; osmotic pressures; 2:2b and c solution equilibria; 2:3 phase diagrams and data; 2:5a viscosity and diffusion; 2:5b thermal conductivity; gas and solid phase reaction rates: 2:7 electric conductivity of aqueous and non-aqueous solutions; emf; dissociation constants in aqueous solution; 2:8 optical properties of glass and other substances; 2:9 and 2:10 magnetic properties, including some nmr data. Primary sources are cited throughout. Beginning in 1961, the "new series" of Landolt-Börnstein Tables (Ref. 12.10) began appearing.

12.12 *Lange's Handbook of Chemistry.* 11th ed. Edited by John A. Dean. New York: McGraw-Hill, 1973. 1570 pp.

- "Lange," like the *CRC Handbook* (Ref. 12.4), represents an attempt to distill the most frequently sought data from the vast literature of chemistry and chemical technology. In format and coverage, the two handbooks resemble each other, so it is reasonable to compare them. It must be noted immediately that there is a lot of overlap of coverage, but even for those classes of data that are covered by both, each will usually include some substances that the other doesn't.

Lange has lengthy tables of physical properties of inorganic and organic compounds, which include color, crystalline form, refractive index, density, melting and boiling points, and an indication of solubility. Beilstein references accompany most of the organic data. The uninverted form of organic compound names is used. A melting point index precedes the organic data table. The CRC includes more organic compounds than Lange, but Lange is particularly notable for its coverage of natural products, and has separate tables of physical properties of alkaloids and glucosides. There are also special tables for commercial organic compounds.

Lange has separate sections for analytical chemistry (Section 5) and electrochemistry (Section 6). These areas are scattered in the *CRC Handbook*, and appear to be less well covered there. Section 8 in Lange on spectroscopy is actually devoted only to x-ray and to atomic emission and absorption spectroscopy, but there is extensive coverage of these areas. Among the other unique, useful tables is one which gives the specific gravities of aqueous solutions of common inorganic salts at various concentrations.

12.13 *The Matheson Unabridged Gas Data Book. A Compilation of Physical and Thermodynamic Properties of Gases.* Edited by William Braker and Allen L. Mossman. East Rutherford, NJ: Matheson Gas Products, 1974. 4 vol.

• Physical properties, such as critical constants, refractive index, spectral properties, densities, viscosities, solubilities, thermodynamic constants and compressiblity factors, are tabulated for some seventy commercially important gases. Properties that are important for the safe handling of these gases, including flammability limits in air and auto-ignition temperatures, are also given.

The fifth edition of the *Matheson Gas Data Book*, published by the same company in 1971, includes more gases, but presents less data. It is, however, more oriented towards day-to-day practical considerations, and includes for each gas short sections on precautions in handling and storage, leak detection, cylinder and valve description, and similar topics.

12.14   National Research Council. *International Critical Tables of Numerical Data, Physics, Chemistry, and Technology.* New York: McGraw-Hill, 1926-1930. 7 vol., plus index.

• The data presented were critically evaluated, with references to the original literature. In many instances more up-to-date values are available. However, the *International Critical Tables* are still often consulted because they are familiar and easy to use, and because for certain kinds of physical properties values measured in the 1920s are acceptable for many purposes today.

When using the older literature, it may be important to be cognizant of the numerical values used by the authors. In Volume One, there is a table of variations in atomic weight values used betwen 1882 and 1925. The one volume index contains entries for the individual chemical substances that are found in the tables of properties. This is an unusual and valuable feature, in that most compilations of this sort are indexed by property, but not by substance.

12.15   Rossini, Frederick D. *Fundamental Measures and Constants for Science and Technology.* Cleveland: CRC Press, 1974. 132 pp.

• The author, one of the leading authorities on scientific data, discusses in detail the measurement of very fundamental physical entities, especially length, mass, time, temperature, pressure, atomic weights, and energy. Many useful tables have been culled from other sources, such as those which specify the temperature differences between the 1948 and 1968 temperature scales. Chapter 13 can serve as a small guide to the past and present literature of numerical data in science and technology.

12.16   *Tables Internationales de Constantes Selectionées.* Oxford: Pergamon, 1947-1970. 17 vol.

• Each volume is a separate monograph of tables on a topic of current interest at the time it was published, with emphasis on rapidly growing fields. For some of the volumes, the data, which were taken from the literature, were critically evaluated. The texts are either entirely in French, or both French and English. Five of the volumes (6, 9, 10, 11 and 14) treat optical rotatory power, but Volume 6 is superseded by 14. Two important sources of data are Volume 8 (1958) for oxidation-reduction potentials, and Volume 17 (1970) for spectroscopic data on diatomic molecules.

12.17   U.S. National Bureau of Standards. Office of Standard Reference Data. *National Standard Reference Data Series.* Washington: U.S. Government Printing Office, 1964-.

• This is one of the most authoritative series for physical and chemical properties data. The careful, critical evaluation of data by experts is a costly and time-consuming enterprise, and hence, only a small portion of the properties and substances of interest can be found in this series. Fields of interest that are within the scope of the *National Standard Reference Data Series* (NSRDS) include nuclear properties, atomic and molecular properties, thermodynamic and transport properties, solid state properties, chemical kinetics, and colloid and surface properties.

NSRDS-NBS-55 is a property index to the series, and to other publications for which the Office of Standard Reference Data provided some support, covering 1964-1972. It is updated annually in the Journal of Physical and Chemical Reference Data (Ref. 12.8). The titles in the series, as well as other sources of critically evaluated data, are listed in the *CRC Handbook of Chemistry and Physics* (Ref. 12.4). No attempt is made here to list all of the important topics covered in the series. However, some of the most important publications are treated immediately below in separate entries. For a more complete indication of the data of chemical interest in the series, the user may use the sources which were mentioned above, or the U.S. National Bureau of Standards *Publications of the NBS* which annually indexes the open-literature publications emanating from the Bureau.

12.17.1 NSRDS-NBS-3. *Selected Tables of Atomic Spectra, Atomic Energy Levels and Multiplet Tables.* By Charlotte E. Moore. (1965-). In sections published at irregular intervals.

● Sections are produced in response to the appearance of appropriate data in the literature. By 1976, seven sections had appeared, and included data on silicon, carbon, nitrogen, hydrogen, and oxygen.

12.17.2 NSRDS-NBS-10. *Selected Values of Electric Dipole Moments for Molecules in the Gas Phase.* By Ralph D. Nelson, David R. Lide, Jr., and Arthur A. Mayott. (1967). 49 pp.

● Dipole moments are listed for over 500 molecules.

12.17.3 NSRDS-NBS-24. *Theoretical Mean Activity Coefficients of Strong Electrolytes in Aqueous Solutions from O to 100ºC.* By Walter J. Hamer. (1968). 271 pp.

● The activity coefficients are recorded as a function of ionic strength and temperature, as calculated by seven different equations based on models of interionic attraction. These tables are useful for estimating activity coefficients when there are no experimental values available, and also as an aid in reading the literature where use is made of these models.

12.17.4 NSRDS-NBS-31. *Bond Dissociation Energies in Simple Molecules.* By B. deB. Darwent. (1970).

● Except for organic compounds containing only one carbon atom, the data are limited to inorganic compounds.

12.17.5 NSRDS-NBS-34. *Ionization Potentials and Ionization Limits Derived from the Analyses of Optical Spectra.* By Charlotte E. Moore. (1970). 8 pp., plus 8 foldout tables.

● Critically evaluated first and higher ionization potentials for nearly all the atoms are given. Updated and more complete data for the lanthanides and actinides were published by Martin et al. in J. Phys. Chem. Ref. Data, 3:771-779 (1974).

12.17.6 NSRDS-NBS-35. *Atomic Energy Levels as Derived from the Analyses of Optical Spectra.* By Charlotte E. Moore. (1971). 3 vol.

• The overall arrangement is by increasing atomic number. The energy levels for the different atomic states are given, and in those cases for which the spectrum was observed in a magnetic field, g-values are also included. The tables for each element are preceded by a short discussion, and list of important references. Reader and Sugar have published more up-to-date data for iron (J. Phys. Chem. Ref. Data, 4:353-440 (1975)), and Sugar and Corliss have done the same for chromium (J. Phys. Chem. Ref. Data 6:317-383 (1977)). Moore has included atomic energy levels in NSRDS-NDS-3 (Ref. 12.17.1), some sections of which have appeared since 1971.

12.17.7 NSRDS-NBS-37. *JANAF Thermochemical Tables.* 2d ed. By R.D. Stull and H. Prophet (1971).

• The tables give the enthalpies, entropies, and Gibbs energies of formation for over 1100 monatomic and polyatomic species, mainly inorganic. Large molecules are, in general, not included. Supplements have been published in J. Phys. Chem. Ref. Data 3:311-480 (1974), and 4:1-175 (1975). These tables are considered to be among the most important and authoritative sources of thermochemical data.

12.17.8 NSRDS-NBS-39. *Tables of Vibrational Frequencies, Consolidated Volume I.* By T. Shimanouchi. (1972). 160 pp.

• This volume extends, revises, and consolidates numbers 6, 11 and 17, and includes data on the fundamental vibrational frequencies of 223 molecules. Shimanouchi has supplemented this with data on 212 additional molecules in "Tables of Molecular Vibrational Frequencies Consolidated Volume II," published in the *Journal of Physical and Chemical Reference Data*, Vol. 6, p. 993-1102 (1977). An empirical formula index to both volumes appears at the end of Volume II.

13
Specialized Data Compilations

ABSORPTION AND EMISSION OF
ELECTROMAGNETIC RADIATION—SPECTRA

The quantum mechanical atomic and molecular models point to a very close relationship between atomic and molecular structural features, and absorption and emission spectra. Absorption spectra result because the absorption of electromagnetic radiation, when it is passed through a sample, is a function of the radiation frequency. Emission spectra result when a sample has been excited, usually by heat. Absorption spectra are much more routinely measured in chemistry laboratories than emission spectra.

Physical chemists and theoreticians have interpreted spectra to prove the mysteries of atoms and molecules, and their interactions. The quantum mechanical interpretation relates the radiation absorption or emission frequencies to atomic and molecular energy levels, and theoreticians have found it very useful to separate molecular energies into components, with electronic, vibrational and rotational energy components being of principal interest in chemistry.

Because absorption and emission spectra are very characteristic of chemical species, and because intensities at specific frequencies can be accurately correlated with concentrations of species, spectroscopic techniques are important for qualitative and quantitative analyses. Different instrumental techniques have developed for different frequency (or wavelength) regions of the spectrum. Thus, analytical chemists refer to ultraviolet (UV), infrared (IR) or microwave spectra, whereas physical chemists will speak of electronic, vibrational or rotational spectra. While there is a very rough correspondence of electronic with UV, vibrational with IR, and rotational with microwave, the associations are by no means exclusive.

Reference books that deal with spectra may define their scope in terms of either classification, or some combination of both. Hence, index entries are provided in this guide for both points of view.

Techniques based on magnetic resonance and other phenomena also produce "spectra." Reference books dealing primarily with these other techniques are listed elsewhere in this chapter. However, Raman spectroscopy is considered within the scope of this section. Works that treat a variety of "spectroscopy" topics are included in Chapter 12. Consult the index for complete guidance to a specific topic, such as "IR spectra."

Following this section of collections in which the spectra themselves are reproduced are works in which the data from spectra are recorded.

13.1.1   American Petroleum Institute. Research Project 44. *Selected Raman Spectral Data*. College Station, TX: Thermodynamics Research Center and American Petroleum Institute, 1948-. Multivolume.

● The looseleaf format allows updating sheets to be inserted. Spectra are reproduced, and spectral data are presented, with emphasis on hydrocarbons. A companion publication with the same title, issued by the Thermodynamics Research Center, emphasizes compounds other than hydrocarbons.

13.1.2   Bentley, Freeman F., Lee D. Smithson, and Adele L. Rozek. *Infrared Spectra and Characteristic Frequencies.* ~ *700−300* cm^{-1}. New York: Interscience, 1968. 779 pp.

● The low frequency region of the infrared spectrum covered in this book is not the one commonly utilized by organic chemists. Sometimes referred to as the far infrared region, the authors prefer to call it the cesium bromide region, to reflect the use of cesium bromide prisms as the radiation dispersing element in the instruments used at the time the book was published. Chapter 13, "Summary of Characteristic Frequencies," is an extensive collection of frequency-structure correlation tables. Chapter 15 is a collection of over 1500 spectra.

13.1.3   *DMS UV Atlas of Organic Compounds*. New York: Plenum, 1966-1971. 5 vol.

● This is a collection of evaluated UV and visible absorption spectra of organic compounds in solution, along with a few inorganic compounds of interest to the organic chemist. Some tables showing the effects of substituents and solvents are also included. Spectra are grouped according to compound type, with a formula and name index in Volume 5. Each page is devoted to a single spectrum, which is presented in larger size than is customary. Experimental details of the measurement, literature references, and other data such as melting point are also given.

13.1.4 Kirschenbaum, Donald M. (editor). *Atlas of Protein Spectra in the Ultraviolet and Visible Regions.* New York: IFI/Plenum, 1972. 2 vol.

• Examples are given to illustrate the subtle differences among the spectra of proteins. In addition to the general index, there is an index of sources and one of additives and modifiers.

13.1.5 Lang, Laszlo (editor). *Absorption Spectra in the Ultra-Violet and Visible Region.* New York: Academic, 1961-. Multivolume.

• Large, well drawn spectra are presented and on facing pages are the corresponding wavelength/intensity data. For some compounds, the data are given for more than one solvent. All experimental conditions are recorded in each case. In the past, indexes were provided as separate booklets, but beginning with Volume 21 (1977), the indexes are bound in. Volume 21 is the first bound volume (in contrast to the earlier loose-leaf format) and it is the first volume under the imprint of a new publisher for the set—Robert E. Krieger of Huntington, New York.

13.1.6 Mecke, R. *Infrared Spectra of Selected Chemical Compounds.* London: Heyden and Son, 1965-1970. 8 vol.

• The infrared spectra of nearly 2000 compounds, mainly organic, are provided in the 700 cm^{-1} to 5000 cm^{-1} range. The emphasis is on compounds of general interest. Small libraries which cannot afford the more comprehensive Sadtler collection (Ref. 13.1.9) may find Mecke sufficient for many of their needs. For the student who needs to find the IR spectrum of a relatively common compound, this collection is easy to use and may be the best source to try first. Indexes by name and formula are in the last volume, with separate listings for deuterated compounds, polymers and inorganics.

13.1.7   Nyquist, Richard A., and Ronald O. Kagel. *Infrared Spectra of Inorganic Compounds (3800-45 cm⁻¹.* New York: Academic, 1971. 495 pp.

- The nearly 900 spectra are mainly of salts, oxides and some coordination compounds. They were measured with mulls, using inert windows—i.e., windows which did not exchange ions with the samples.

13.1.8   Pouchert, Charles J. *The Aldrich Library of Infrared Spectra.* 2d ed. Milwaukee, WI: Aldrich Chemical Co., 1975. 1571 pp.

- Over 10,000 rather small reproductions of spectra have been compressed into this bulky volume. The spectra were usually recorded either for the pure compounds or for nujol mulls. Because the spectra are so small, the peak positions cannot be read with much precision. Spectra are arranged by chemical class, and there are indexes by name and formula.

13.1.9   *Sadtler Standard Grating Spectra.* Philadephia: Sadtler Research Laboratories, 1966-. Multivolume.

- Together with the prism spectra (Ref. 13.1.10) this is the most comprehensive commercially available infrared collection. By 1977, there were almost 53,000 grating spectra published in 53 loose-leaf binders. At present all spectra are prepared at the Sadtler Research Laboratories, though in the past some were accepted from other sources. The loose-leaf format allows the company to send those libraries which have holdings of the spectra updated sheets when they become available.

  These spectra are indexed in the *Total Spectra Index* (see Ref. 13.1.13). There is, in addition, an index which is specific to the grating spectra—the Grating Spec-Finder. The Spec-Finder can lead to the identity of an unknown compound on the basis of a spectrum obtained in the laboratory. The user merely identifies the strongest band within his spectrum, and the strongest bands within the sub-regions of the spectrum, locates which spectra match those specifications *via* the Spec-Finder, and then looks the matching spectra up in the Sadtler collection. Further help in matching is provided by the Chemical Class Codes which accompany each entry in the Spec-Finder. Details about how to use the Spec-Finder can be found at the beginning of the 1976 cumulative volume.

13.1.10 *Sadtler Standard Prism Spectra.* Philadelphia: Sadtler Research Laboratories, 1956-. Multivolume.

• Grating and prism spectra result from slightly different instrumentation for producing infrared absorption spectra. The difference depends on whether a salt crystal prism or a diffraction grating is used to disperse the infrared radiation into a continuous spectrum of frequencies. The recent trend has favored grating over prism instrumentation. There are over 50,000 infrared prism spectra in the collection. Spectra for some compounds can be found in both the grating and prism collection. In addition to those spectra which have been measured with both kinds of instrumentation, some prism spectra have been reissued as part of the grating collection. They are numbered in the grating collection from 38001P to 48487P. The prism spectra are indexed in the *Total Spectra Index* (Ref. 13.1.13), and there is a Prism Spec-Finder similar to the Grating Spec-Finder (see Ref. 13.1.9).

13.1.11 *Sadtler Standard Raman Spectra.* Philadelphia: Sadtler Research Laboratories.

• There are over 4000 spectra in this collection.

13.1.12 *Sadtler Standard Ultraviolet Spectra.* Philadelphia: Sadtler Research Laboratories, 1961-. Multivolume.

• This very extensive collection of over 40,000 spectra is indexed in the *Total Spectra Index* (Ref. 13.1.13). The well-drawn reproductions of the spectra are each accompanied by tables giving the molecular extinction coefficients at different wavelengths, and the conditions of measurement.

13.1.13 Sadtler Research Laboratories. *Total Spectra Index.* Philadelphia. 1972-. Multivolume.

• Indexes are by name, formula, chemical class and Sadtler spectral number, and they are to the Sadtler grating and prism infrared spectra (Ref. 13.1.9 and 13.1.10), the Sadtler 60 MHz NMR spectra (Ref. 13.19.2), the Sadtler UV spectra (Ref. 13.1.12), and a collection of Sadtler DTA (differential thermal analysis) thermograms. A few other collections are also indexed, including the Coblentz infrared spectra.

The compound names correspond to those used by *Chemical Abstracts* before 1972. There are also some cross references from common names. The Chemical Class Index presents a hierarchical array based on structural features. There are nearly 100 codes, up to three of which may be assigned to a given compound, plus special codes indicating if the compound is alicyclic, aliphatic, aromatic, etc. Thus, the Chemical Class Index groups together similar compounds (and of course, as any compound classification scheme must, it separates compounds whose similarities do not match the priorities of this particular scheme).

A cumulation covering all years of the spectra was published in 1976, and supplements to that cumulation are now being issued.

13.1.14   Sadtler Research Laboratories. *Using the Sadtler Spectra.* Philadelphia: 1 cassette, 4-track, plus 30 color slides, 2 x 2 in.

• This audio-visual aid for learning how to use the Sadtler spectra is especially valuable in connection with the Chemical Class Index and the Spec-Finders.

13.1.15   Schrader, B., and W. Meier. *DMS Raman/IR Atlas.* Weinheim: Verlag Chemie, 1975. 2 vol.

• For each of the approximately 1000 compounds included, the IR and Raman spectra are illustrated facing each other. While most of the IR spectra can easily be found in other sources, there are fewer alternative collections of Raman spectra. The spectra are particularly well presented in this *Raman/IR Atlas.* Since IR and Raman spectra compliment each other in the interpretation of molecular vibrations, this *Atlas* is particularly suitable for demonstrating the constrast between the two for a large variety of compounds.

13.1.16   Szymanski, Herman A. *Interpreted Infrared Spectra.* New York: Plenum, 1964-1967. 3 vol.

• The spectra in this book are ideally presented for instructing the user in the interpretation of infrared spectra. Group frequency bands of model compounds are shaded for emphasis, and their associated vibrational modes identified. The spectra themselves are grouped together in various chemical classes, and the author has included concise discussions of the notable features of each class. Vibrational analysis data and correlation tables are also given.

13.1.17 Welti, David. *Infrared Vapour Spectra*. London: Heyden, 1970. 211 pp.

• Nearly all infrared collections present spectra taken from samples in the condensed phases. However, vapor phase spectra are noteworthy because the influences of neighboring molecules are much smaller than in the condensed phases. Welti offers over 300 spectra, primarily of compounds which are liquids at room temperature, but vapors at the conditions which are indicated with the spectra. The arrangement is by chemical class, preceded by a formula index. Following the spectra, there is a formula index to vapor and gas phase spectra found in major commercial collections.

ABSORPTION AND EMISSION OF
ELECTROMAGNETIC RADIATION. DATA.

The researcher, technician, or student who obtains a spectrum in the laboratory will usually prefer to match that with a good graphic reproduction in a spectral collection. However, a table of data will do if the graphic reproduction is not available. In fact, for very precise quantitative analysis, or for a theoretical interpretation depending on numerical values, a critical, authoritative table is preferred.

Correlation tables, which are particularly useful to the student who wants to learn how to interpret spectra are also included here. Correlation tables in effect summarize vast amounts of observed data on different compounds, to indicate which spectral regions are associated with specific structural features, and also to indicate how absorption or emission frequencies might be shifted by factors such as neighboring group or solvent influences.

13.2.1 Bashkin, Stanley, and John O. Stoner, Jr. *Atomic Energy Levels and Grotrian Diagrams*. Amsterdam: North-Holland, 1975-. Multivolume.

• Volume 1. Hydrogen I—Phosphorus XV. Volume 2. Sulfur I—Calcium XX. Atomic energy levels are obtained

from spectroscopic studies and are of fundamental significance to many areas of chemistry. The data in this set, which
have been derived from other secondary publications as well
as original research papers, are presented graphically. The
clarity and detail of presentation are noteworthy.

13.2.2   Beck, R., W. Englisch, and K. Gürs. *Table of Laser Lines in
          Gases and Vapors.* Berlin: Springer, 1976. 130 pp.

● Lasers are becoming increasingly important to chemists
as sources for the emission of electromagnetic radiation,
with applications as diverse as Raman spectroscopy and
isotope separation. This book is a catalog of over 4000 laser
lines from atomic and molecular media, with references to
the original literature. As there is no index, the user must
scan the table of contents to locate substances of interest as
emitters.

13.2.3   Dolphin, David, and Alexander Wick. *Tabulation of Infrared
          Spectral Data.* New York: Wiley, 1977. 549 pp.

● The detailed interpretation of infrared spectra requires the
imposition of steric, electronic and solute-solvent interaction
factors on the group frequency correlations. This reference
book serves as a guide to those factors, by tabulating data for
a large number of model compounds. Each chapter deals with
a different type of structure or functional group, and begins
with an interpretive discussion. In a separate section at the
end of the book, the spectra of about 60 common organic
solvents are reproduced.

13.2.4   Hirayama, Kenzo. *Handbook of Ultraviolet and Visible
          Absorption Spectra of Organic Compounds.* New York:
          Plenum, 1967. 642 pp.

● The wavelengths of maximum absorption and the extinction coefficients are given for a large number of organic
molecules. Table 1 is arranged by chromophore (the table of
contents should be consulted), while Table 2 is arranged by
increasing absorption maximum and can serve as an index to
Table 1. Becuase UV and visible spectra are determined by
the chromophores present in molecules, this is a valuable aid
in interpreting those spectra.

13.2.5 International Union of Pure and Applied Chemistry. Commission on Molecular Structure and Spectroscopy. *Tables of Wavenumbers for the Calibration of Infrared Spectrometers.* Edited by A.R.H. Cole. 2d ed. Oxford: Pergamon, 1976. 219 pp.

• Part I provides the standards for the instruments of moderately high resolution which are needed for physical studies of molecular structure and dynamics. Part II has standards for the lower resolution instruments which are usually used by organic and inorganic chemists to aid in identifying compounds. In both parts, a series of vaporphase spectra is displayed, accompanied by tables identifying the positions of some of the sharp peaks in the spectra.

13.2.6 Kaufman, Victor, and Bengt Edlen. J. Phys. Chem. Ref Data, 3:825 895 (1974). "Reference Wavelengths from Atomic Spectra in the Range 15 Å to 25000 Å."

13.2.7 Massachusetts Institute of Technology. *Wavelength Tables.* Cambridge, MA: M.I.T. Press, 1969. 429 pp.

• Title on page vii: Tables of wavelengths and intensities of the principal atomic spectrum lines in the range 10,000 — 2000 Å.
The data are arranged by decreasing wavelength, and for each wavelength, the emitting atom is given, along with its stage of ionization, the intensity of the line, and a literature reference. The data have been critically evaluated and are held in high repute.

13.2.8 *Microwave Spectral Tables.* Washington: U.S. Government Printing Office, 1964-1968. 5 vol.

• National Bureau of Standards Monograph 70. Volume 1 Diatomic Molecules. Volume 2 Line Strengths of Asymmetric Rotors. Volume 3 Polyatomic Molecules with Internal Rotation. Volume 4 Polyatomic Molecules without Internal Rotation. Volume 5 Spectral Line Listing. Other related data, such as dipole moments, are given in many cases. Some judgment was made with regard to the accuracy of the data to be included. Volumes 1, 3 and 4 have separate tables for individual molecules, in which the spectral frequencies are reported and interpreted. A revision of Volume 1 was published by Frank Lovas and Eberhard Tiemann, J. Phys. Chem. Ref. Data, 3:609-769 (1974). Volume 5, designed as a ready reference, lists the lines that were reported in Volumes 1, 3 and 4 in ascending order of frequency.

13.2.9   Miller, Roy. *Irscot: Infrared Structural Correlation Tables.* London: Heyden, 1964-1973. 11 vol., plus index.

- Tables: 1. Hydrocarbons. 2. Halogen Compounds. 3. Oxygen Compounds excluding acids. 4. Carboxylic acids and Derivatives. 5. Nitrogen Compounds excluding N-O compounds. 6. N-O Compounds. 7. Heterocyclics. 8. Sulphur Compounds. 9. Silicon Compounds. 10. Boron Compounds. 11. Phosophorous Compounds.

Absorption band positions, both in wavenumbers and in wavelength, are given for a selection of compounds for each mode of vibration. The format and arrangement, as well as the separately bound indexes, make these tables very easy to use.

13.2.10   *Organic Electronic Spectral Data.* New York: Interscience, 1946/1952-. Annual.

- Each volume is arranged by molecular formula, and for each compound the solvent, wavelengths of maximum absorption, and literature reference are given. Annual publication began in 1966. Prior to that, each volume covered more than one year. The dates on the spines of the volume refer to the years covered, not the date of publication, and there is a lag of about six years between the period covered and the publication.

13.2.11   Parsons, M.L., B.W. Smith and G.E. Bentley. *Handbook of Flame Spectroscopy.* New York: Plenum, 1975. 478 pp.

- Chapter 2 is an atlas of atomic absorption lines, from page 70 to 152, atomic emission lines from page 153 to 242, and atomic fluorescence lines from page 243 to 285. Chapter 3, "Fundamental Information," is devoted to the theoretical models used to explain flame spectra, and includes many useful tables. There are data on atomic energy levels and transition probabilities, and the percent of atoms in the lower energy levels for different elements at various temperatures. Other related data, such as bond dissociation energies, are also included.

13.2.12 Pestemer, Max. *Correlation Tables for the Structural Determination of Organic Compounds by Ultraviolet Light Absorptiometry.* Weinheim: Verlag Chemie, 1974. 157 pp.

• In his Introduction, Pestemer points out that ultraviolet spectra do not lend themselves as readily to spectra-structure correlation tables as do the infrared spectra. However, ultraviolet bands are particularly sensitive to influences on π-electron systems, and hence, the arrangement in this book is based upon the degree of unsaturation. For the most part, the author permits the user to supply most of the interpretation. The data themselves are taken from major secondary compilations.

13.2.13 Shimanouchi, T. *Tables of Vibration Frequencies.* (See Ref. 12.17.8.)

13.2.14 Suchard, S.N. (editor). *Spectroscopic Data.* New York: IFI/Plenum, 1975-1976. 2 vol. in 3.

• Volume 1. Heteronuclear Diatomic Molecules. Volume 2. Homonuclear Diatomic Molecules (by Suchard and J.E. Melzer).
Electronic spectral data are obtained from absorption spectra and from other techniques. In addition to presenting data that describe the electronic spectra of selected molecules this reference work also provides constants related to the electronic states of those molecules, such as energy levels, equilibrium internuclear distances, and dissociation energies.

13.2.15 Szymanski, Herman A., and Ronald E. Erickson. *Infrared Band Handbook.* 2d ed. New York: IFI/Plenum, 1970. 2 vol.

• Volume 1: 4240 cm^{-1} to 999 cm^{-1}. Volume 2: 999 cm^{-1} to 29 cm^{-1}, and formula index.
For each frequency, the structural formula of the absorbing molecule is given, along with a qualitative indication of its intensity, the conditions under which the measurement was made, and a reference to the original literature. In some

cases, the assigned mode of vibration is also indicated. the bands are included if they are the principal IR parameters for identifying a compound.

13.2.16   Varsanyi, G. *Assignments for Vibrational Spectra of Seven Hundred Benzene Derivatives.* New York: Wiley, 1974. 2 vol.

● This book begins with a discussion of the normal modes of vibration of benzene and its derivatives and a table of characteristic frequencies of substituents. The bulk of Volume 1 is taken up with tables for individual molecules in which the vibration frequencies are correlated with normal modes. Spectra are reproduced in Volume 2, which concludes with an index to both volumes.

BIOCHEMISTRY

Because of the vastness of the field of biochemistry, there is no attempt in this guide to cover its reference works, for which a separate guide would be appropriate. However, the general interest that many chemists have in biological chemistry cannot be ignored, and for that reason, as well as the need the chemist has for data on natural products, important biochemical data collections are listed here.

13.3.1   *Atlas of Protein Sequence and Structure.* Volume 5. Edited by Margaret O. Dayhoff. Silver Spring, MD: National Biomedical Research Foundation, 1972-. Irregular.

● Sequences of proteins and nucleic acids are tabulated, based on a critical review of the literature. Coverage is comprehensive. Calculated molecular weights and other data are also included. Volume 5 supersedes earlier volumes, and is updated by supplements which were published in 1973 and 1976.

13.3.2   Barrell, B.G., and B.F.C. Clark. *Handbook of Nucleic Acid Sequences.* Oxford, England: Joynson-Bruvvers, 1975. 104 pp.

● One of the most exciting areas of modern scientific research is the study of the various RNAs and DNAs and their role in heredity and other life processes. Techniques for establishing the sequence of the nucleotides in nucleic acids are relatively new, the first sequence having been published in 1965. Over 100 sequences are graphically displayed in this handbook, which is notable for its clarity of presentation. References to the original literature are included.

13.3.3   Devon, T.K., and A.I. Scott. *Handbook of Naturally Occurring Compounds.* New York: Academic, 1972-. Multivolume.

● Volume 1: Acetogenins, Shikimates and Carbohydrates. Volume 2: Terpenes.
For each compound, a box is drawn containing the name, structural formula, molecular weight and formula, optical rotation, melting point, a literature reference, and a classification number used to specify the structural type. Each volume has indexes by name, molecular formula and molecular weight.

13.3.4   *Handbook of Biochemistry and Molecular Biology.* 3rd ed. Edited by Gerald D. Fasman. Cleveland: CRC Press, 1975-. Multivolume.

● Volume 1: Proteins; Volume 2: Nucleic Acids; Volume 3: Lipids, Carbohydrates, Steroids; Volume 4: Miscellaneous Physical and Chemical Data. Oxidation-reduction potentials, thermodynamic quantities, molecular parameters, spectroscopic data, and much else, including a great deal of information about the composition and structure of specific proteins and nucleic acids is included. The volumes also contain sections on nomenclature.

13.3.5   National Research Council. Committee on Biological Chemistry. *Specifications and Criteria for Biochemical Compounds.* 3rd ed. Washington: National Academy of Sciences, 1972. 216 pp.

● "This publication is the result of a program to improve the quality of chemicals available for biochemical research by establishing criteria, standards, or specifications useful for describing such chemicals, particularly with regard to purity" (from the Preface). Methods of purification and of assaying for purity are also often included.

13.3.6   *Natural Product Chemistry.* Edited by Koji Nakanishi, et al. Tokyo: Kodansha, 1974-1975. 2 vol.

● Chapter 1 presents four different classification schemes for natural products. Spectroscopic and related data are presented in Chapter 2. Most of the remaining chapters deal with specific classes of natural products. The editors make extensive use of graphics to illustrate structural features and reactions.

13.3.7   Yamaguchi, K. *Spectral Data of Natural Products.* Amsterdam: Elsevier, 1970. 765 pp.

● Natural products are classified into twenty groups based on structure. Data are given from those techniques which have been particularly useful in elucidating structures, including UV, IR, NMR, optical rotatory dispersion, circular dichroism, and mass spectrometry. Melting points and other physical constants are also given. Structures are drawn, and some spectra are reproduced.

CHROMATOGRAPHY

Chromatography is a laboratory separation technique, capable in many cases of achieving a very fine division of a chemical mixture. There are various kinds of chromatography, depending on the media employed. The range of chemical types and mixture compositions that can be successfully submitted to chromatographic separation is enormous, and hence, this technique is found in nearly every type of chemical laboratory in operation.

There are many chemists and chemical technicians who use chromatography in a routine manner, but who are not themselves experts on chromatography. The ability to obtain data useful for effecting specific separation is thus of great benefit to a large number of laboratory practitioners.

13.4.1   American Society for Testing and Materials. *Compilation of Gas Chromatographic Data.* Philadelphia: 1967. 732 pp.

● (ASTM Data Series DS 25A). *First Supplement.* Philadelphia: 1971. 726 pp. (AMD 25 A-S1). "The object of this compilation is to provide the gas chromatographer with the gas chromatographic conditions used for the separation of a wide variety of compounds and to facilitate the tentative identification of materials producing peaks on chromatograms" (page iv). It is possible to find data and literature references useful for the separation of a specific compound (from others) on the basis of its molecular formula. Data are also tabulated on the basis of the liquid phase, and this can be used in the tentative identification of unknown materials.

13.4.2   American Society for Testing and Materials. *Liquid Chromatography Data Compilation*. Philadelphia: 1975. 186 pp. (Atomic and Molecular Data Series AMD 41).

● If an experimenter wants to separate a certain compound from another by liquid chromatography, he might consult these tables for relevant data. All of the information has been taken from the primary literature, and abstracts of that literature are included.

13.4.3   Zweig, Gunter and Joseph Sherma. *Handbook of Chromatography*. Cleveland: CRC Press, 1972. 2 vol.

● Retention ratios, retention times, retention volumes, and other data, with literature citations, are presented in Volume 1. This volume concludes with an index of the over 12,000 compounds that are included in the tables. The second volume is "designed to give the researcher . . . a working knowledge of the theory and practices . . . " (from the Preface). It also includes a section on detection methods for paper and thin-layer chromatography, descriptions of methods of sample preparation, and a bibliography of books, arranged by publisher.

CONVERSION TABLES

Tables for converting from one unit to another can be found in many handbooks and general reference works, including the *CRC Handbook of Chemistry and Physics* (Ref. 12.4), *The Chemist's Companion* by Gordon and Ford (Ref. 12.5), and *Lange's Handbook of Chemistry* (Ref. 12.12). Most academic libraries will have a collection of dictionaries and tables devoted exclusively to unit conversion. One example of a particularly extensive reference work of this sort is included here. In the Library of Congress Classification, many of these books are between QC 82 and QC 96.

13.5.1   Chiu, Yishu. *A Dictionary for Unit Conversion*. Washington: School of Engineering and Applied Science, The George Washington University, 1975. 451 pp.

• This is a particularly extensive listing of conversion factors for length, area, volume, mass, weight, force, time, or their combinations.

13.5.2   Coleman, Charles DeWitt, William R. Bozman, and William F. Meggars. *Table of Wavenumbers*. Washington: U.S. Government Printing Office, 1960. 2 vol.

• National Bureau of Standards Monograph 3. The table is for converting wavelengths in air to wavenumbers in vacuum, a conversion which is important for precise interpretation of chemical spectra in terms of molecular or atomic energy levels.

CRYSTALLINE SOLIDS

Historically, crystalline solids have had a special importance in chemistry because one of the most extensively used techniques for getting a pure compound has been crystallization. The molecules, atoms or ions in a crystal are arranged in highly consistent patterns. This has permitted the development of experimental techniques for probing the detailed structure of the microsopic world in crystals. Although all such studies have not depended on the diffraction of x-rays, that technique has been most highly developed, and is responsible for the existence of structural data dating back to the early part of the twentieth century. X-ray diffraction has been greatly refined since its early days, and is probably still the single most important technique for complete molecular structure determination. Applications toward the identification of materials are also important.

13.6.1   *Atlas of Steroid Structure*. Edited by William L. Duax and Dorita A. Norton. New York: IFI/Plenum, 1975-. Multi volume.

• Detailed structural data are given, accompanied by drawings illustrating the major structural features. It is the intention of the editors to present the crystallographic data in a manner that can be interpreted by the non-crystallographer biochemist. In Volume 1, 184 steroids for which structures were reported between 1945 and 1974 are included.

13.6.2 Donnay, Joseph D.H. and Helen M. Ondik. *Crystal Data Determinative Tables*. 3rd ed. Washington: U.S. Department of Commerce, National Bureau of Standards, and the Joint Committee on Powder Diffraction Standards, 1972. 2 vol.

• Volume 1. Organic compounds; Volume 2. Inorganic compounds.

Cell dimensions, space group, number of formula units per cell, density, and in some cases, other data are presented, based on x-ray, electron and neutron diffraction experiments. Each volume has indexes by formula, chemical name and mineral name. There are also valuable appendices, such as a concordance of space-group notations. A companion to this publication (Ref. 13.6.4) has appeared in the *Journal of Physical and Chemical Reference Data*.

13.6.3 Joint Committee on Powder Diffraction Standards. Associateship at the National Bureau of Standards. *Powder Diffraction Data*. Swarthmore, PA: 1976. 440 p. and *Search Manual for Powder Diffraction Data*. Swarthmore, PA: 1976. 152 pp.

• The diffraction pattern that results from x-rays being scattered by a sample of a crystalline solid in powdered form is characteristic of that solid and can be used in its identification. Approximately 900 compounds, mainly inorganics, are included in *Powder Diffraction Data*. The Search Manual serves as an index to that data based on the eight most intense lines of each substance. Hence, these two volumes in combination are useful for the identification of a relatively common crystalline substance from its x-ray powder diffraction pattern.

13.6.4 Mighell, Alan D., Helen M. Ondik, and Bettijoyce Breen Molino, J. Phys. Chem. Ref. Data, 6:675-829 (1977). "Crystal Data Space Group Tables."

• The authors describe this publication as a companion to the *Crystal Data Determinative Tables* (Ref. 13.6.2). All of the compounds which were treated in the Determinative

Tables are listed here (by formula only) according to the space group of which they are members. Within each space group, the compounds are ordered by cell dimension.

13.6.5   *Structure Reports.* Utrecht: Bonn, Scheltema, and Holkema, 1913/28-. Annual.

• The publisher and title have varied. The first seven volumes, under the title *Strukturbericht*, were published by Akademische Verlagsgesellschaft, in Leipzig. All subsequent volumes are in English and were published for the International Union of Crystallography.

These are very careful, critical reports on structure determinations that have been reported in the literature, including many relatively obscure sources. There is, at present, usually a lag of two or three years before all the volumes covering a given year are complete. The data derived principally from x-ray diffraction studies, but also from other techniques, cover metals, and inorganic compounds (Part A) and organic compounds (Part B). For many of the reports that are given, particularly for structures of less than thirty atoms, sufficient *structural* information is given so that there is no need to consult the original paper if only data are needed. Unit cell dimensions and symmetry information are given for each structure, and in many cases, the unit cells are illustrated. Other structurally significant data, such as interatomic distances, are also often included. The short discussions which accompany each report offer further clarification of the structures and often relate them to other known structures. Organic compounds that have been studied by vapor phase electron diffraction are treated briefly at the end of the organic sections.

Parts A and B each have their own subject, formula and author indexes. Sixty-year cumulative indexes, covering 1913-1973, have been published for both parts.

13.6.6   Wyckoff, Ralph. *Crystal Structures.* 2d ed. New York: Interscience, 1963-1971. 6 vol. in 7.

• An attempt is made to present all crystal structures in which all or most of the atom positions have been determined. This reference work is noted for the excellent illustrations, which accompany the discussions and tabulations of data.

ELECTROCHEMISTRY

Atoms and molecules may acquire electric charges in various ways. A neutral molecule may have an uneven distribution of charge (i.e., a dipole moment) or may be polarized by an electric field. These processes, as well as the behavior of the charged species in an electric field, are of fundamental importance in chemistry and provide the basis for many important technological applications.

For explanations of the analytical and research techniques based on electrochemistry, and some of the more important electrochemically based technologies, the reader is advised to consult *The Encyclopedia of Electrochemistry* by Clifford A. Hampel (Ref. 10.24).

13.7.1   Conway, B.E. *Electrochemical Data*. Amsterdam: Elsevier, 1952. 374 pp.

   • Reprinted by Greenwood Press, Westport, CT in 1969.
   Tables present data on dielectric constants, dipole moments, activity coefficients, transport properties of strong electrolytes in solution, dissociation constants, solubilities and solubility products, electrochemistry of melts at high temperatures, oxidation-reduction potentials, and electrode properties. The author included only data of known accuracy; literature references are amply provided.

13.7.2   *Digest of Literature on Dielectrics*. Washington: National Academy of Sciences, 1936-. Annual.

   • The annual issues contain tables of dielectric constants, dipole moments, and dielectric relaxation times, as well as articles reviewing the literature and extensive bibliographies.

13.7.3   Dobos, D. *Electrochemical Data. A Handbook for Electrochemists in Industry and Universities*. Amsterdam: Elsevier, 1975. 339 pp.

   • Originally published in Hungarian (Budapest: Technical Publishing House, 1965). SI units are used in the English edition.
   This book is intended as a rapid source for practical information. Included are conductivities and related data in aqueous and non-aqueous solutions of electrolytes, and in the melt, activity coefficients, oxidation-reduction potentials, and electrokinetic, polarographic and related data.

Two additional useful features are a section on electrochemical equations and formulae, and a bibliography of
texts and reference works.

13.7.4   Meites, Louis, and Petr Zuman. *Electrochemical Data*. New
York: Wiley, 1974-. Multivolume.

• The compiler-editors of this exemplary reference set are
both leading workers in the field of electrochemistry. Part I
of *Electrochemical Data* is restricted to organic, organometallic and biochemical substances. Inorganics will be treated in
Part II. Individual volumes are limited to the coverage of
specific kinds of substances, and to specific time periods.

The emphasis is on polarography and related areas; con-
ductometry, potentiometry, the various electrochemical ti-
tration techniques, and some other topics are excluded. Data
which does not meet the editors' standards has either been
excluded, or included with qualifying statements.

13.7.5   Milazzo, Guilio, and Sergio Caroli, *Tables of Standard
Electrode Potentials*. New York: Wiley-Interscience, 1978.
419 pp.

• This is a very extensive tabulation of data. It was not
critically evaluated.

EQUILIBRIUM IN SOLUTION

A chemical system achieves equilibrium when it shows no further
tendency to change its properties. The equilibrium constant is a
measure of the degree to which a reaction is complete at equilibrium,
and it is definable in terms of the thermodynamic free energy.

There is a vast literature on equilibrium in solution because of its
importance in analytical and other applications, and the interpretation
of the properties of living systems.

13.8.1   Chemical Society (London). *Stability Constants of Metal-
Ion Complexes*. 2d ed. London: 1964. 2 vol. (Special Publication No. 17). *Supplement 1*. 1971. 865 pp. (Special Publication No. 25). Compiled by L.G. Sillen and A.E. Martell.

• These Chemical Society publications provide fairly comprehensive data on the dissociation of metal complexes, as

well as references to the original literature. Both organic and inorganic ligands are included.

13.8.2 Kortüm, G., W. Vogel and K. Andrussow. *Dissociation Constants of Organic Acids in Aqueous Solution*. London: Butterworths, 1961. 558 pp.

- Reprinted from Pure Appl. Chem., *1*:187-536 (1961). This collection of dissociation constants for over 1000 organic acids is similar in format to the other IUPAC-sponsored compilations of this type of data (Ref. 13.8.5 and Ref. 13.8.6).

13.8.3 Marcus, Y., and D.G. Howery. *Ion Exchange Equilibrium Constants*. London: Butterworths, 1975. 41 pp.

- Ion exchange is important both as a laboratory technique for many analytical preparative procedures, and as a facet of some commercial processes. Hence, this short compilation, prepared under the sponsorship of the IUPAC Commission on Equilibrium, is very valuable.

13.8.4 Martell, Arthur E., and Robert M. Smith, *Critical Stability Constants*. New York: Plenum.

- Volume 1: Amino Acids (1974) 469 pp. Volume 2: Amines. (1975) 415 pp. Volume 3: Other Organic Ligands. (1977) 495 pp. Volume 4: Inorganic Complexes. (1976) 257 pp. Equilibrium constants, and the enthalpies and entropies of complexation have been critically evaluated for presentation in this reference work. Complexes in which the ligands are coordinated to protons as well as both transition and nontransition metal ions have been included. This treatment is to be contrasted with the Chemical Society tables (Ref. 13.8.1) which include a greater variety of ligands, but which present unevaluated data.

13.8.5 Perrin, D.D. *Dissociation Constants of Inorganic Acids and Bases in Aqueous Solutions*. London: Butterworths, 1969.

- Reprinted from Pure Appl. Chem. *20*:133-236 (1969). Over 200 substances, including many hydrated metal ions as well as the more conventional inorganic acids and bases, such as hydrochloric acid and ammonia, are included. For each substance, the pK values are recorded at one or more temperatures, with information about the experimental conditions and references to the original literature. Conflicting values from different sources are often included.

13.8.6   Perrin, D.D. *Dissociation Constants of Organic Bases in Aqueous Solution*. London: Butterworths, 1965. 515 pp.

• Dissociation constants of 3800 compounds are recorded, many of them at several different temperatures. The preparation of this table was sponsored by IUPAC. A 235 page Supplement published in 1972 (London: Butterworths) extends coverage to many more compounds. The Supplement includes an index that covers both itself and the original work.

INORGANIC CHEMISTRY

A lot of the most basic physical data for inorganic compounds can be found in the *CRC Handbook of Chemistry and Physics* (Ref. 12.4) and similar sources described in Chapter 12. However, the user will often have to look elsewhere for specialized information. Gmelin's handbook (Ref. 14.5) aims for comprehensive coverage of inorganic chemistry data, but it is up to date for only a small percentage of inorganic substances.

Monographs dealing with specific elements and the most common inorganic compounds abound, and they often provide data not otherwise easily found (i.e., requiring a detailed search of the primary literature). In the Library of Congress system, QD 181 is reserved for books dealing with specific elements and their compounds. Books within this class are arranged alphabetically by element symbol (e.g. QD 181.A3 for books on silver; QD 181.B2 for books on barium; QD 181.B4 for books on beryllium.)

13.9.1   Ball, M.C., and A.H. Norbury. *Physical Data for Inorganic Chemists*. London: Longman, 1974. 175 pp.

• The data were collected together in this book for the use of students in inorganic chemistry courses. Most of it has been taken from other secondary sources. Ionization potentials, electron affinities, thermodynamic data, lattice energies, electrode potentials, bond dissociation energies, and similar basic data suitable for solving problems in undergraduate inorganic chemistry are included.

13.9.2 Christensen, James J., Delbert J. Eatough and Reed M. Izatt. *Handbook of Metal Ligand Heats.* 2d ed. New York: Dekker, 1975. 495 pp.

• Enthalpy, entropy, heat capacity and equilibrium constant data are compiled for over one thousand ligands,. The ligands determine the arrangement of the tables, and there are indexes by formula, synonym, and central element.

13.9.3 *Handbook of Solid-Liquid Equilibria in Systems of Anhydrous Inorganic Salts.* Edited by N.K. Voskresenskaya. Jerusalem: Israel Program for Scientific Translations, 1970. 2 vol.

• Originally published in Russian in 1961. Phase data, and some phase diagrams, are presented for binary and multicomponent systems.

13.9.4 König, E., and S. Kremer. *Ligand Field Energy Diagrams.* New York: Plenum, 1977. 454 pp.

• Nearly 400 energy diagrams for d-electrons in various symmetries are drawn.

13.9.5 Wells, A.F. *Structural Inorganic Chemistry.* 3rd ed. Oxford: Clarendon, 1962. 1055 pp.

• This very thorough monograph is included here because of the many clear illustrations of crystal structures of elements, alloys and inorganic compounds.

KINETICS

Chemical kinetics is the study of reaction rates. Chemists study rates primarily to determine reaction mechanisms. An excellent overview of the field, with guidance to the original literature, is provided by the massive treatise *Comprehensive Chemical Kinetics* (Ref. 21.2). The sources listed here can be used for a quick look up of rate constants or other kinetic data for many, but not all, of the reactions which have been studied.

13.10.1   Denisov. E.T. *Liquid-Phase Reaction Rate Constants.* New York: IFI/Plenum, 1974. 771 pp.

  • Translated from Russian (Moscow: Nauka, 1971). Rate constants are presented for reactions in the liquid phase which in nearly all cases involve free radicals either as reactants, products, or intermediates. Chapter 8 treats ionic oxidation-reduction reactions. There is no index.

13.10.2   Kondratiev, V.N. *Rate Constants of Gas Phase Reactions.* Springfield, VA: National Technical Information Service, 1972. 428 pp.

  • Most of the information in this book was originally published in Russian. The translation was published by the Office of Standard Reference Data of the U.S. National Bureau of Standards, and distributed by the National Technical Information Service (NTIS), with NTIS report number COM-72-10014. The literature on the kinetics of bimolecular and termolecular reactions of neutral species, to the end of 1969, is covered, and much of the data have been critically evaluated. An extensive table of equilibrium constants has been appended by the translators.

13.10.3   United States. National Bureau of Standards. *Tables of Chemical Kinetics: Homogeneous Reactions.* Washington: U.S. Government Printing Office, 1951. 731 pp. (NBS Circular No. 510)

  • Supplement No. 1 to Circular 510 was issued in 1956, and Supplement No. 2 in 1960. Subsequently, supplementary tables were issued as National Bureau of Standards Monograph 34, Volume 1 in 1961 and Volume 2 in 1964. Rate constants, activation energies and Arrhenius frequency factors are tabulated for a large number of reactions in the gas or solution phases. All the data have been critically evaluated, and references to the original sources are given. The arrangement, which is by reaction type, may be understood by scanning. Supplement 2 is an alphabetical index by compound type to Circular 510 and its first supplement. It also includes an index by reaction type to the classification scheme. These tables are products of the Chemical Kinetics Data Project at the National Bureau of Standards. This same group has also been responsible for two critical reviews which were published as part of the NSRDS series (see Ref. 12.17): NSRDS-NBS 9, *Tables of Bimolecular Gas Reactions* by A.F. Trotman-Dickenson and G.S. Milne (1967), and NSRDS-NBS 20, *Gas Phase Reaction Kinetics of Neutral Oxygen Species*, by H.S. Johnston (1968).

LIQUIDS AND SOLUTIONS

Although the liquid phase is less well understood than the crystalline or gas phases, it is easier to work with. Consequently, most of the chemistry in school and university course-related laboratories, and a great deal of the chemistry in research and technology, is carried out in liquid-phase solution. There is a massive amount of empirical data on the properties of pure liquids and liquid-phase solutions. Many important properties such as vapor pressures and viscosities are not easily located in the primary literature. Hence, tabulations, such as those listed here are most welcome.

13.11.1 Boublík, Tomáš, Vojtěch Fried, and Eduard Hála. *The Vapour Pressures of Pure Substances*. Amsterdam: Elsevier, 1973. 626 pp.

• Subtitle: Selected values of the temperature dependence of the vapour pressures of some pure substances in the normal and low pressure region.
Experimental and calculated vapor pressures are presented for over seven hundred pure compounds (mostly organic) at various temperatures.

13.11.2 Frier, Rolf K. *Aqueous Solutions*. Berlin: Walter de Gruyter, 1976-. Multivolume.

• Solubilities, usually over a range of temperatures, redox potentials, dissociation constants, and thermodynamic data for aqueous solutions of inorganic and organic compounds are given.

13.11.3   Hirata, Mitsuho, Shuzo Ohe, and Kunio Nagahama. *Computer Aided Data Book of Vapor-Liquid Equilibria*. Tokyo: Kodansha, 1975. 933 pp., plus index.

• With the aid of some experimental data, and model equations, vapor pressure and related data are calculated and tabulated for about 1000 binary mixtures.

13.11.4   Horsely, Lee H. *Azeotropic Data—III*. Washington: American Chemical Society, 1973. 628 pp. (Advances in Chemistry Series No. 116).

• Supersedes Advances in Chemistry Series Nos. 6 and 35. Data are presented for binary, ternary, and a few higher systems. If an azeotrope forms, the boiling points of the components and of the azeotrope, as well as the azeotropic composition are provided. Mention is also made of some mixtures which form no azeotrope. The data have not been evaluated, but if there have been appreciable differences from the same investigator, only the most recent data have been included.

13.11.5   Linke, William F. *Solubilities. Inorganic and Metal-Organic Compounds*. 4th ed. Princeton, NJ: D. Van Nostrand, 1958-1965. 2 vol.

• The data are derived from the literature and presented in tables which illustrate solubilities in water at various temperatures, and in the presence of various concentrations of other substances. Some solubilities in solvents other than water are also included.

13.11.6   Marsden, Cyril (editor). *Solvents Guide*. 2d ed. New York: Interscience, 1963. 633 pp.

• Physical properties, such as vapor pressure, solubility in water, boiling point, viscosity, etc. are given for solvents of importance in chemical technology. In addition, for each solvent a list of its azeotropic mixtures with other liquids, along with the boiling points of these mixtures, is provided, as well as information on physiological properties, industrial grades, storage, and handling. The appendices include a number of useful conversion tables and a table of toxic hazards due to solvent vapors.

13.11.7 Mellan, Ibert. *Industrial Solvents Handbook.* 2d ed. Park Ridge, NJ: Noyes Data Corp., 1977. 567 pp.

• Physical and chemical properties, such as specific gravity, flash point, viscosity, freezing and boiling point, and many others, are given for the pure and mixed solvents which are used in industry.

13.11.8 *Solubilities of Inorganic and Organic Compounds.* Edited by H. Stephen and T. Stephen. New York: Macmillan, 1963-1964. 2 vol. in 4.

• This work is based on and extends a Russian publication by V.V. Kafarov (Moscow, 1962). Binary systems are in Volume 1, ternary and multi-component systems in Volume 2. Emphasis is on the more common laboratory reagents and solvents, including water, but data are given over ranges of temperatures. Each volume has its own index.

13.11.9 Timmermans, Jean. *The Physico-Chemical Constants of Binary Systems in Concentrated Solutions.* New York: Interscience, 1959-1960. 4 vol.

• Volume 1. Two Organic Compounds (without Hydroxyl Derivatives). Volume 2. Two Organic Compounds (at least one a Hydroxyl Derivative). Volume 3. Systems with Metallic Compounds. Volume 4. Systems with Inorganic + Organic or Inorganic Compounds (excepting Metallic Derivatives).

Boiling points, freezing points, densities, critical solution temperatures, heats of solution, dilution, etc., viscosities, and other properties are given at various compositions.

MASS SPECTRA

A mass spectrum results when a beam of ions is separated into a "spectrum" based on mass-to-charge ratios. The ions may be single atoms or atom clusters. Precision instrumentation has made possible many applications of this technique. Chemical analyses and molecular structure determinations by mass spectrometry have particularly been facilitated by the publication of data compilations such as those listed here.

13.12.1   American Petroleum Institute. Research Project 44. *Selected Mass Spectral Data*. College Station, TX: Thermophysics Research Center and American Petroleum Institute, 1961?-. Multivolume.

- Each table gives the data for a specific compound. Many of the spectra were recorded considerably earlier than 1961. The ASTM *Index of Mass Spectral Data* (Ref. 13.12.2) may serve as an index. Stenhagen's *Registry of Mass Spectral Data* (Ref. 13.12.6) is more comprehensive, but it does not actually provide numbers for the mass-to-charge ratios or the intensities, as this collection of data does.

13.12.2   American Society for Testing and Materials. Committee E-14 on Mass Spectrometry. *Index of Mass Spectral Data. Listed by Molecular Weight and the Six Strongest Peaks*. Philadelphia: 1969. 624 pp. (ASTM Publication AMD 11).

- The arrangement which allows the user to approach the index by molecular weight or by the mass-to-charge ratio of the six strongest peaks is designed for the experimenter who needs to identify a compound on the basis of its mass spectrum. The compound would, of course, have to be among the approximately 8000 relatively well known organic compounds that are included. Unfortunately, there is no index by name or formula in this volume. The spectral data were supplied from several sources. One of those sources was the American Petroleum Institute Research Project 44, and hence, this volume may serve as an index to their publication(see Ref. 13.12.1).

13.12.3   Binks, R., J.S. Littler, and R.L. Cleaver. *Tables for Use in High Resolution Mass Spectrometry*. London: Heyden and Son, 1970. 160 pp. Accompanied by: D. Henneberg and K. Casper. *Chemical Formulae from Mass Determinations*. 25 pp.

- These tables, along with the accompanying booklet, can be used to identify the elemental composition of species causing the individual peaks in a mass spectrum for organic compounds containing H, N, O, S, Si, Cl, Br, B or F atoms. A number of other useful tables, such as those illustrating the peak patterns to be expected for certain combinations of heteroatoms, are also valuable for helping to identify unknown compounds.

13.12.4 Cornu, A., and R. Massot. *Compilation of Mass Spectral Data*. 2d ed. London: Heyden, 1975. 2 vol.

- The data from over 10,000 spectra are arranged in three sections: A by molecular weight; B by molecular formula; and C by fragment ion values. In each part, the ten most intense peaks of each spectrum are given. As there is some overlap in source spectral collections between this and the ASTM (American Society for Testing and Materials) Index (Ref. 13.12.2), many of the same compounds are covered by both, but the arrangements are different.

13.12.5 *Eight Peak Index of Mass Spectra*. Aldermaston, Reading, England: Mass Spectrometry Data Centre, 1970. 2 vol.

- This is one of the most extensive collections of mass spectral data for the identification of compounds. It can be approached either by molecular weight or by the mass-to-charge ratios of the peaks.

13.12.6 Stenhagen, Einar, Sixten Abrahamsson, and Fred W. McLafferty. *Registry of Mass Spectral Data*. New York: Wiley, 1974. 4 vol.

- Small bar graphs of 18,806 spectra are arranged by molecular weight and composition, and indexed by formula. The *Registry* is more comprehensive than the *Atlas* which was prepared by Stenhagen et. al. five years earlier. Most compilations of mass spectral data are designed for the identification of a compound from its mass spectrum, but this *Registry* is primarily intended to be used when someone wants mass spectral data for a particular compound. The Preface points out the availability of computer search capabilities for the *Registry*.

MOLECULAR DIMENSIONS AND SHAPES

The most detailed molecular structural information is obtained from x-ray crystallography. Hence, accurate internuclear distances and interbond angles are reported in several of the reference books pertaining to crystalline solids (Ref. 13.6.1 to Ref. 13.6.6). However, x-ray crystallography is not the only technique for obtaining structural parameters. Parameters obtained from spectroscopic measurements can be found in Suchard's *Spectroscopic Data* (Ref. 13.2.14). Sections of the Landolt-Börnstein Tables (Ref. 12.10) include parameters based on several techniques, as do the tables from the Chemical Society of London, listed here.

13.13.1 *Tables of Interatomic Distances and Configuration in Molecules and Ions.* London: Chemical Society, 1958. (Special Publication No. 11) and *Interatomic Distances Supplement.* 1965. (Special Publication No. 18).

● Both volumes have similar arrangements: Section S., a table of selected bond lengths; Section M which contains the bond lengths and interbond angles for specific inorganic and organic compounds; Section R, the bibliography; and Section C, a classified index of organic compounds. The data were obtained by a variety of methods, including rotational spectra and x-ray and other diffractional methods. Though old, much of it is still considered reliable today.

13.13.2 *Molecular Structures and Dimensions. Series A. Interatomic Distances.* Edited by Olga Kennard et al. Utrecht: N.V.A. Oosthoek's Uitgevers Mij Utrecht, 1972-. Irregular.

● Issued as part of the bibliography *Molecular Structures and Dimensions* (Ref. 4.16), Series A is intended as a continuation of the Chemical Society's Special Publications 18 and 11 (Ref. 13.13.1).

NUCLEAR DATA OF INTEREST IN CHEMISTRY

13.14.1 Fuller, Gladys H. J. Phys. Chem. Ref. Data, 5:835-1092 (1976). "Nuclear Spins and Moments."

● The nuclear spins and moments are of interest to chemists working with a number of important techniques including microwave spectroscopy and nuclear magnetic resonance.

ORGANIC CHEMISTRY

Most of the entries in the Chemical Substance Index of *Chemical Abstracts* are organic compounds. Since, in effect, most chemical compounds are organic compounds, most data compilations deal either entirely or in part with organic chemistry. The reader may therefore be puzzled by the inclusion of an "Organic Chemistry" section.

In a way, this is a miscellaneous, or catch-all section, held together only by the requirement that each item be devoted to some aspect of organic chemistry. However, most of the resources listed in this section will be of particular value to the laboratory chemist engaged in the more traditional pursuits of experimental organic chemistry.

13.15.1   American Petroleum Institute. Research Project 44. *Selected Values of Properties of Hydrocarbons and Related Compounds.* College Station, TX: Thermodynamics Research Center and American Petroleum Institute, 1942.-. Multi-volume.

• These tables have been issued by a number of different agencies, and came under the auspices of the API Project 44 in 1961. The loose-leaf format allows replacement with revised tables and the filing of new tables in their proper places. Boiling and freezing points, refractivities, densities, surface tensions, vapor pressures, critical properties and thermodynamic properties are tabulated. A companion set with the title *Selected Values of Properties of Chemical Compounds* is issued by the Thermodynamics Research Center of College Station, Texas.

13.15.2   Dewar, Michael J.S., and Richard Jones. *Computer Compilation of Molecular Weights and Percentage Compositions for Organic Compounds.* Oxford: Pergamon, 1969. 476 pp.

• The compound must have carbon and hydrogen, and may have up to two different heteroatom elements, selected from among the following: bromine, chlorine, fluorine, iodine, nitrogen, oxygen, or phosphorus. Since elemental compositions are routinely performed in the identification of organic compounds, this book and the one by Ege (Ref. 13.15.3) are convenient time-savers for the laboratory chemist who would otherwise have to calculate the percentages himself. Arrangement is based on empirical formula, with no index.

13.15.3   Ege, Günter. *Elementary Analysis Tables*. Weinheim/ Berg-str.: Verlag Chemie, 1966. 355 pp.

● (Title is also given in German and French.) Ege's tables may be used more directly than Dewar's (Ref. 13.15.2) in determining an empirical formula based on percent composition, but they are limited to compounds containing carbon, hydrogen, nitrogen and either oxygen or sulfur. Ege has arranged his tables primarily by increasing percent carbon and secondarily by increasing percent hydrogen, so that if an experimenter has a compound with a certain percent composition, he can look up all of the empirical formulas that match.

13.15.4   Jordan, T. Earl. *Vapor Pressure of Organic Compounds*. New York: Interscience, 1954. 266 pp.

● The author intended comprehensive coverage of the technologically important organic compounds at that time, and consequently data on some 1500 or more compounds are given. The data are presented in the form of tables, and graphs of vapor pressure versus temperature.

13.15.5   Klyne, W., and J. Buckingham. *Atlas of Stereochemistry*. *Absolute Configurations of Organic Molecules*. London: Chapman and Hall, 1974. 311 pp.

● "The purpose of this book is to bring together in a readily accessible form a proportion of the vast mass of data which exists in the literature concerning the absolute configurations of chiral molecules" (from the author's Introduction). The absolute stereochemical configurations of model compounds are provided, along with references to the original literature. Key compounds from the main groups of natural products are among those compounds included.

13.15.6   Rappoport, Zvi. *Handbook of Tables for Organic Compound Identification*. 3rd ed. Cleveland: Chemical Rubber Company, 1967. 564 pp.

● This is a standard reference work that is a very useful aid in identifying organic compounds. In the main table, compounds are organized in classes according to increasing boiling points. Data are given for the compounds, and in many cases, also for their derivatives. There are also a few other tables, such as IR correlation charts, and a table of the mutual miscibility of organic solvents, and there is an extensive organic compound index.

13.15.7 Timmermans, J. *Physico-Chemical Constants of Pure Organic Compounds.* Amsterdam: Elsevier, 1950-1965. 2 vol.

• Volume 1 (1950) and Volume 2 (1965) are arranged in the same way, but Volume 2 provides new data, as well as some corrections to the data in Volume 1. Basic physical data for specific compounds are tabulated, including: vapor pressure at various temperatures, viscosity, critical constants, density, specific heat, etc. Each volume has an index of compounds and a list of references. The literature between 1910 and 1957 is covered, but only the more common laboratory reagents are included.

13.15.8 Utermark, Walther, and Walter Schicke. *Melting Point Tables of Organic Compounds.* 2d ed. New York: Interscience, 1963. 715 pp.

• The approximately 7000 compounds are arranged by increasing melting point, and for each one, the structure, some physical constants including boiling point, density, etc., some reactions with common reagents, and a Beilstein reference are given. This table complements Rappoport (Ref. 13.15.6) which is based on boiling points, and is useful in the identification of organic compounds.

13.15.9 Verschueren, Karel. *Handbook of Environmental Data on Organic Chemicals.* New York: Van Nostrand Reinhold, 1977. 659 pp.

• Data on air pollution, water pollution, and biological effects on microorganisms, plants, animals, and man for over a thousand organic compounds are presented.

13.15.10 Wilen, Samuel H. *Tables of Resolving Agents and Optical Resolutions.* Notre Dame, IN: University of Notre Dame Press, 1972. 308 pp.
• Tables in Chapter III present resolving agents and chiral absorbents for classes of optically active compounds. In

Chapter IV data are given for some 1200 specific organic compounds for which resolutions have been reported in the literature. The author cautions that, although his tabulation is substantial and comprehensive, all of the compounds which have been resolved are not listed, but that essentially all classes of organic compounds that have been resolved by conventional methods are represented.

13.15.11 Wilhoit, R.C., and B.J. Zwolinski. *Physical and Thermodynamic Properties of Aliphatic Alcohols.* New York: Published by the American Chemical Society and the American Institute of Physics for the National Bureau of Standards, 1973. 420 pp. (J. Phys. Chem. Ref. Data, Vol. 2, Suppl. 1).

- "This monograph represents the most exhaustive review and critical analysis of certain physical and thermodynamic properties of aliphatic alcohols that has been published in the world literature of chemistry during the last 100 years" (from the Preface). The literature up to 1967 is covered. Refractive indexes, vapor pressures, boiling points, densities, critical properties, heat capacities, entropies and enthalpies for phase changes, and similar data, are discussed and tabulated.

ORGANOMETALLIC CHEMISTRY

Organometallic compounds are claimed by both the inorganic and organic chemistry communities. This guide chooses not to take sides in that debate. Listed here are three sources of organometallic data that will be useful to all chemists who work with these materials.

13.16.1 Burger, Kalman. *Organic Reagents in Metal Analysis.* Oxford, England: Pergamon, 1973. 266 pp.

- Table 27 (Page 140-175) is an alphabetical list of reagents indicating which metal ions they may be used to determine, and by which method. Table 28 (Page 176-184) is the reverse table, alphabetical by the metal ion. Subsequent tables give additional data for analytical determinations, including: spectrophotometric parameters, thermal stabilities, gravimetric factors, stability constants of the chelates of complexing agents used as analytical reagents, and solvent data. The thirty-eight most important reagents are discussed in detail in the first part of the book.

13.16.2 *Handbook of Organometallic Compounds.* Edited by Nobue Hagihara, Makoto Kumada and Rokuro Okawara. New York: Benjamin, 1968. 1044 pp.

 • Basic information, including a method of preparation and physical properties such as melting points and refractive indexes are given for a large number of organometallic compounds. Primary literature references are included.

13.16.3 Kaufman, Herbert C. *Handbook of Organometallic Compounds.* Princeton, NJ: D. Van Nostrand, 1961. 1546 pp.

 • The arrangement is based on the periodic table. Basic information for each compound is given, including color, specific gravity, refractive index, melting point, heat of formation, vapor pressure, etc., depending upon what information was available in the literature at the time this book was prepared.

PHOTOCHEMISTRY

The induction of chemical reactions by light, and the production of light by chemical reactions constitute the subject matter of photochemistry. Usually, the subject is considered to be limited to visible and ultraviolet radiation, and excludes ordinary emission and absorption spectroscopy.

13.17.1 Murov, Steven L., *Handbook of Photochemistry.* New York: Dekker, 1973, 272 pp.

 • The various parameters that are important in photochemical research are presented in the twenty-six sections of this book. Some spectroscopic data are also included. Preceding the subject and formula indexes at the end of the book, there is a guide to photochemical literature.

POLYMERS

The high molecular weight species, referred to as polymers, or occasionally as giant molecules, are of major technological importance. Plastics, rubbers, synthetic fibers, silicone greases, and many other modern "miracles" of technology are basically polymers. The molecules of life are also, for the most part, polymers, and the reader is referred to Ref. 13.3.2, Ref. 13.3.4 and other biochemical reference works for data on them.

The data compilations listed here emphasize commercial polymers. *The Encyclopedia of Polymer Science and Technology* (Ref. 10.35) is another valuable source of data, and provides references to additional publications. There is not sufficient room in this guide to include mention of the several important reference works in materials science which have polymer data in them.

13.18.1   Brandrup, J., and E.H. Immergut (editors). *Polymer Handbook.* 2d ed. New York: Wiley, 1975. 1363 pp.

> • The chapters in this important reference work are authored by different specialists. In spite of the opening section on nomenclature and units, there are many inconsistencies in the nomenclature used in the book, and systematic nomenclature is sometimes abandoned in favor of trivial names. The subject index is a broad index of properties. A few of the property entries are subdivided by substance, but in general the index cannot be used in connection with questions about specific polymers. The fifth section is divided into fourteen chapters, each one on the physical constants of an important polymer. Other sections deal with polymerization and depolymerization, properties in the solid state and in solution, and the physical properties of monomers and solvents, as well as a few minor subjects.

13.18.2   *Modern Plastics Encyclopedia.* New York: McGraw-Hill, 1941-. Annual.

> • The title has varied, and there have been gaps in the publication sequence.
>
> There are three sections. The first, for general information, most closely resembles an ordinary encyclopedia with short articles on specific polymers and other subjects. The Engineering Data Bank (the second section) is an extensive table of specification data on equipment and on plastics materials. The third section is a buyers' guide. The encyclopedia is issued as part of a subscription to the magazine *Modern Plastics.*

RESONANCE SPECTRA

In resonance spectroscopy an external field is used to populate a normally near-empty energy state, in order to be able to induce transitions between the excited and normal states by passing electromagnetic radiation through the system. For an appropriate design of the conditions, a spectrum will result which may be interpreted in terms of molecular structure.

Organic chemists have especially developed the technique of proton magnetic resonance—wherein the external field is magnetic and conditions are such that the environments of protons in molecules can be studied—to the point where it is a standard, almost routine, technique in the organic laboratory. In fact, the phrase "nuclear magnetic resonance" is often used to mean "proton magnetic resonance," although the former is more general, embracing techniques for studying other nuclei. Most of the reference books listed here are for proton magnetic resonance but other resonance spectra and data are included as well.

13.19.1 *Sadtler Standard Carbon-13 NMR Spectra.* Philadelphia: Sadtler Research Laboratories, 1976-. Multivolume.

> • Two thousand spectra are included within the first ten volumes. Volume 10 (1977) includes a cumulative numerical index, based on the spectrum numbers in the carbon-13 NMR collection, which provides cross references to the Sadtler prism (Ref. 13.1.10), grating (Ref. 13.1.9), UV (Ref. 13.1.12), and nmr (Ref. 13.19.2) spectra. There are also cumulative name, formula and chemical class indexes, and a cumulative spec-finder that permits the identification of an unknown compound on the basis of its spectrum.

13.19.2 *Sadtler Standard Nuclear Magnetic Resonance Spectra.* Philadelphia: Sadtler Research Laboratories, 1967-. Multivolume.

> • By 1977 there were 26,000 spectra in this collection, making this the most comprehensive commercial catalog of nmr spectra available. Chemical shifts (relative to tetramethylsilane) are cited, and are considered to be accurate to

±0.05 ppm. Calculated coupling constants are sometimes provided for some of the less usual spin-spin coupling patterns. The *Total Spectra Index* (Ref 13.1.13) serves as an index. The nmr colection has a 1976 cumulative index by increasing molecular weight, which also gives the IR and UV spectrum numbers when applicable. A particularly useful set of cumulative indexes based on the chemical shifts was published in 1976, which will aid the user both in interpreting spectra and in identifying unknowns from their spectra. An index arranged by a code based on the nature of the proton and its immediate surroundings was also published cumulatively in 1976.

13.19.3   Simons, W.W. *The Sadtler Guide to NMR Spectra.* Philadelphia: Sadtler Research Laboratories, 1972. 542 pp.

● Four hundred and eighty spectra, selected from the *Sadtler Standard NMR Spectra* (Ref. 13.19.2), are used to illustrate the principles of proton nuclear magnetic resonance. Arranged in a manner similar to many organic chemistry textbooks, this guide presents characteristic spectra for various proton types, with brief interpretations accompanying each spectrum. There is an index by name, and some useful appendices, including one that lists compounds by increasing chemical shift.

13.19.4   Simons, W.W., and M. Zanger. *The Sadtler Guide to the NMR Spectra of Polymers.* Philadelphia: Sadtler Research Laboratories, 1973. 298 pp.

● The book was designed to facilitate the use of nmr spectra in characterizing specific polymers and also in analyzing copolymers and polymer mixtures. The format is similar to Ref. 13.19.3. In addition to spectra of polymers, spectra of appropriate monomers and reference compounds are portrayed. There are indexes by chemical and commercial name, and a polymer finder chart that allows for some correlations.

RESONANCE SPECTRAL DATA

13.20.1   Biryukov, I.P., M.G. Voronkov, and I.A. Safin. *Tables of Nuclear Quadrupole Resonance Frequencies.* Jerusalem: Israel Program for Scientific Translations, 1969. 135 pp.

- Originally published in Russian (Leningrad: Izdatel'stvo "Khimiya," 1968). Approximately half of the data are for chlorinated inorganic and organic compounds, reflecting the importance of chlorine in this area of study.

13.20.2 Bovey, Frank A. *NMR Data Tables for Organic Compounds.* New York: Interscience, 1967. 610 pp.

- This extensive table of chemical shifts and coupling constants, with references to the original literature, is arranged by compound formulas.

13.20.3 Breitmaier, E., G. Haas, and W. Voelter. *Atlas of Carbon-13 NMR Data.* London: Heyden, 1975-. Multivolume.

- For each compound, the peak positions are listed, along with experimental conditions and a reference to the literature. Molecular structures are drawn and numbered so that peaks can be identified. Multiplicities are also given. There are indexes by name, chemical class, formula, molecular weight and chemical shift.

13.20.4 Brügel, Werner. *Nuclear Magnetic Resonance Spectra and Chemical Structure.* New York: Academic, 1967. 235 pp.

- Each table presents data for a number of compounds that all illustrate a particular type of structure—e.g., monosubstituted benzenes or vinyl metal compounds. This type of arrangement lends itself well to instruction in the interpretation of nmr spectra. As there is a compound index, the book can also be easily used to obtain data for a specific compound.

13.20.5 Chamberlain, Nugent F. *The Practice of NMR Spectroscopy with Spectra-Structure Correlations for Hydrogen—1.* New York: Plenum, 1974.

- This text is intended to be useful to the experienced spectroscopist but will also be a valuable learning aid for the beginner. There are very extensive correlation tables for both chemical shifts and coupling constants. In addition, 400 typical spectra are reproduced.

13.20.6   Emsley, J.W., and L. Phillips. Progr. Nucl. Mag. Res. Spectr.
          7: 1-526 (1971). "Flourine Chemical Shifts"; Emsley, J.W.,
          L. Phillips, and V. Wray. Progr. Nucl. Mag. Res. Spectr. 10:
          83-756 (1976). "Fluorine Coupling Constants."

          • Both articles are appended with lengthy tables of values
          which were compiled from an extensive review of the litera-
          ture.

THEORETICAL CHEMISTRY

The tables in this section are based on theoretical rather than experimental data. They were all constructed from quantum mechanical
calculations, usually with the aid of high-speed computers.

13.21.1   Fraga, Serafin, Jacek Karwowski, and K.M.S. Saxena. *Hand-
          book of Atomic Data*. Amsterdam: Elsevier, 1965. 551 pp.

          • Energies, electron affinities, ionization potentials, coup-
          ling constants, and other quantities, calculated from the
          Hartree-Fock method, are listed in separate tables.

13.21.2   Krauss, Morris. *Compendium of Ab Initio Calculations of
          Molecular Energies and Properties*. Washington: U.S. Gov-
          ernment Printing Office, 1967. 139 pp. (NBS Technical Note
          438).

          • Best values from *ab initio* calculations reported in the
          literature from 1960 to the time of publication are presented.
          The calculated properties can include total energy, dissocia-
          tion energy, electron affinity, spectroscopic constants, elec-
          tric moments, field gradients, polarizabilities, and magnetic
          moments. Also included are tables of orbital energies.

13.21.3   Miller, James, John M. Gerhauser, and F.A. Matsen. *Quan-
          tum Chemistry Integrals and Tables*. Austin: University of
          Texas Press, 1959. 1223 pp.

          • The tables, which were computed with an IBM 650, are
          limited to one-center and two-center problems (i.e., atoms
          and diatomic molecules.).

13.21.4   Coulson, C.A., and A. Streitwiesser, Jr. *Dictionary of π-
          Electron Calculations*. San Francisco: Freeman, 1965. 358 pp.

- Though much smaller than Streitwiesser's *Supplemental Tables of Molecular Orbital Calculations* (Ref. 13.21.5), the authors refer to this one as the "readable volume" (Preface). Molecular orbital methods were used to calculate, by computer, the properties of π-electrons in simple conjugated and aromatic compounds, including some containing heteroatoms.

13.21.5 Streitwiesser, A., Jr., and J.I. Brauman. *Supplemental Tables of Molecular Orbital Calculations.* Oxford: Pergamon, 1965. 2 vol.

- The *Dictionary of π-Electron Calculations* by Coulson and Streitwiesser (Ref. 13.21.4) is published as a separate section within this much larger work. However, except for the heterosystems included in Coulson and Streitwiesser's Dictionary, this work is limited to hydrocarbons, mainly polynuclear aromatics. Molecular orbital calculations on π-electron systems yielded electron density, π-energy, and related data.

13.21.6 *Table of Molecular Integrals.* 2d ed. By Masao Kotani and others. Tokyo: Maruzen, 1963. 328 pp, plus foldout.

- This collection of formulae and numerical tables is useful for the theoretician who needs to apply sophisticated quantum mechanical techniques towards the solution of many electron problems. These formulae are applicable in principle to many large polyatomic molecules.

THERMODYNAMICS

In the calculations of chemical behavior, the laws of thermodynamics must not be violated. In order to apply these laws, precise values for the thermodynamic properties of substances, especially the enthalpies, entropies, and free energies, are required.

Equilibrium constants, for chemical systems that are truly in a state of chemical equilibrium, can be defined in terms of thermodynamic properties, and indeed free energies and other thermodynamic data are often included in tabulations of equilibrium constants. However, the data on chemical equilibrium in solutions have been deemed worthy of a separate section in this guide, and can be found between Refs. 13.8.1 and 13.8.6.

13.22.1   Barin, Ihsan, and O. Knacke. *Thermochemical Properties of Inorganic Substances*. Berlin: Springer, 1973. 921 pp.

• This extensive tabulation may be used to obtain thermodynamic data for relatively common inorganic substances in either the gaseous or condensed phase. Among the properties included are heat capacity, enthalpy of formation, entropy, and free energy, as a function of temperature, and constants for calculating heat capacity and vapor pressure. Additional information related to phase transitions is often also included. The very short list of literature sources precedes the tables. A supplement issued in 1977 includes updated data on compounds in the initial work, as well as data on new compounds, and also includes a more detailed practical description of how to use the tables.

13.22.2   "CODATA Recommended Key Values for Thermodynamics." *CODATA Bulletin* No. 17 (January 1976) and *CODATA Bulletin* No. 22 (March 1977). Paris: CODATA Secretariat.

• Enthalpy and entropy data for 94 chemical species are presented. Nearly all of the species are monoatomic or diatomic, and hence, the data may be useful in calculating the thermodynamic properties of many other substances.

13.22.3   *Consolidated Index of Selected Property Values. Physical Chemistry and Thermodynamics*. Washington: National Academy of Sciences—National Research Council, 1962. 274 pp.

• This index may serve as a key to the properties reported in the following publications which are included in this guide: Ref. 13.15.1, Ref. 13.22.7, and Ref. 13.22.11 (in part).

13.22.4   Cox, J.D., and G. Pilcher. *Thermochemistry of Organic and Organometallic Compounds*. London: Academic, 1970. 643 pp.

• A very extensive table of enthalpies of formation and reaction is presented in Chapter 5. Hydrogenation and combustion reactions are emphasized, but in some cases other reactions are also included.

13.22.5   *JANAF Thermochemical Tables.* (See Ref. 12.17.7)

13.22.6   Karapet'yants, M. Kh., and M.L. Karapet'yants. *Thermo-dynamic Constants of Inorganic and Organic Compounds.* Ann Arbor: Humphrey Science Publishers, 1970. 461 pp.

• The standard enthalpies of formation, free energies of formation, entropies and heat capacities of over 4000 compounds are arranged in two easy-to-use sequences, one for inorganics, and the other for organics. Values found by approximate calculations, as well as experimental data are included. The authors disclaim any attempt at critical evaluation. This book was originally published in Russian (Moscow, 1968).

13.22.7   Roooini, Frederick D., and others. *Selected Values of Chemical Thermodynamic Properties.* Washington: U.S. Government Printing Office, 1952. 1268 pp. (National Bureau of Standards Circular No. 500).

• The first series of tables gives enthalpies and free energies of formation, entropies and heat capacities at 25°C; the second series gives temperatures, enthalpies, entropies and heat capacities related to phase transitions. Elements, inorganic compounds, and organic compounds with up to two carbon atoms are included.

The first series is being revised, being issued in parts as NBS Technical Note 270 (see Ref. 13.22.9). It should be noted that Circular 500 predates the scale of atomic weights based on carbon-12 which is now accepted by all chemists, and that these new values of atomic weights are used in the revisions.

13.22.8   *Selected Thermochemical Data Compatible with the CO-DATA Recommendations.* By V.B. Parker, et al. Springfield, VA: National Technical Information Service, 1976. 35 pp. (PB-250 845).

• The enthalpy of formation, free energy of formation, entropy, and heat capacity of 384 inorganic substances at 298.15°K are given. The data are based on CODATA recommended values (See Ref. 13.22.2).

13.22.9 *Selected Values of Chemical Thermodynamic Properties.*
Washington: U.S. Government Printing Office, 1965-. Multivolume. (National Bureau of Standards Technical Note 270).

• This updates the Bureau's Circular 500 (see Ref. 13.22.7).
By 1973, seven parts had been issued, and the compounds of
seventy-six of the elements covered.

13.22.10 *Selected Values of the Thermodynamic Properties of Binary
Alloys.* Prepared by Ralph Hultgren et al. Metals Park, OH:
American Society for Metals, 1973. 1435 pp.

• Hultgren had pointed out in an earlier version of this work
*(Selected Values of Thermodynamic Properties of Metals
and Alloys.* Wiley, 1963) that thermodynamic data in the
primary literature are often very erratic, and that the problem of errors in the literature is perhaps more acute for
metals than for most other types of substance. This newer
compilation contains many substantial changes in values
from the older one. The authors point out that there are still
many imperfect measurements reported in the literature. The
data in this publication have been subjected to critical
evaluation by the authors. Enthalpy, entropy, free energy,
and heat capacity data are provided, and some phase diagrams are also included.

13.22.11 U.S. Bureau of Mines. *Bulletin.* Washington: U.S. Government Printing Office.

• There are two series in this Bulletin which provide thermodynamic data: a) Contributions to the Data on Theoretical Metallurgy. *Bulletins* 383, 384, 393, 406, 477, 542, and
584 are indexed in the Consolidated Index (Ref. 13.22.3).
Bulletins 592 and 668 appeared after the Consolidated Index
was published. b) Chemical Thermodynamic Properties of
Hydrocarbons and Related Substances. *Bulletins* 595, 654,
and 666.
Bulletin 401 (The Thermodynamic Properties of Metal
Carbides and Nitrides) is also indexed in the Consolidated
Index.

13.22.12 Vargaftik, N.B. *Tables on the Thermophysical Properties of
Liquids and Gases in Normal and Dissociated States.* 2d ed.
Washington: Hemisphere, 1975. 758 pp.

- This translation of a Russian work is based on the international literature, but includes data from a lot of Russian sources which are not readily available in the West. Molar entropies, enthalpies and volumes, heat capacities, and other data are given for pure organic and inorganic substances, and for some mixtures. In this book, only the commonly encountered materials are treated, but a great deal of data, over wide ranges of temperature and pressure are given for them. For some substances, data on properties not ordinarily expected in tables of thermophysical properties, such as surface tension, are included.

THERMOPHYSICAL PROPERTIES

This section may be used to find data on the properties of materials under the influence of heat. As precise knowledge of these properties is often of considerable technological importance, the critical evaluation of data by experts should be noted.

13.23.1 Ho, C.Y., R.W. Powell, and P.E. Liley. *Thermal Conductivity of the Elements: A Comprehensive Review.* New York: Published by the American Chemical Society and the American Institute of Physics for the National Bureau of Standards, 1975. 796 pp.

- Journal of Physical and Chemical Reference Data, Volume 3, Supplement 1.
  For the 82 elements for which experimental values are available, the authors present that data, with critical comments, and based on their analyses, they give recommended values over a range of temperatures. Estimated values are provided for those elements for which there are no experimental results. An abridged version, which contains only the recommended and estimated values, is also available.

13.23.2 Jamieson, D.T., J.B. Irving and J.S. Tudhope. *Liquid Thermal Conductivity. A Data Survey to 1973.* Edinburgh: Her Majesty's Stationery Office, 1975. 221 pp.

- Organic and inorganic liquids, including molten salts, as well as some solutions, are covered. Liquid metals are excluded. The authors attempt to assess the accuracy of the data.

13.23.3   Purdue University, Lafayette, Indiana. Thermophysical
          Properties Research Center. *Thermophysical Properties of
          Matter*. Edited by Y.S. Touloukian. New York IFI/Plenum,
          1970-. Multivolume.

	A	B	C	D
Thermal Conductivity	V. 1	V. 2	V. 3	
Specific Heat	V. 4	V. 5	V. 6	
Thermal Radiative Properties	V. 7	V. 8		V. 9
Thermal Expansion	V. 12	V. 13		

  A: Metallic elements and alloys
  B: Nonmetallic solids
  C: Nonmetallic liquids and gases
  D: Coatings
           V. 10 Thermal Diffusivity
           V. 11 Viscosity

The data in this collection, drawn from the primary literature, have
been evaluated by experts.

14

Beilstein
and Gmelin

Two of the most important reference works in chemistry are the German productions which are casually known by their shortened names "Beilstein" and "Gmelin". Their full titles are given below. Beilstein covers the research literature of organic chemistry; Gmelin covers inorganic chemistry. They are both excellent sources for the older literature, but, except for certain parts of Gmelin, they are not useful for the most current information. In no way should this diminish the significance of these very important tools, which emphasize thoroughness in their coverage rather than currency. Note that the three entries in the sub-section on aids to using Beilstein are in order of increasing difficulty of use rather than in the customary alphabetical order.

BEILSTEIN

14.1   Beilstein, Friedrich Konrad. *Handbuch der Organischen Chemie.* 4th ed. Berlin: Springer, 1918-. Multivolume.

• It is important to understand how Beilstein is organized. The main work (Hauptwerk) covers the literature of organic chemistry through 1909. Each supplement (Ergänzungswerk) covers a subsequent period: first supplement, 1910-1919; second supplement, 1920-1929; third supplement, 1930-1949; fourth supplement, 1950-1959. The arrangements within the main work and each of the supplements follow similar patterns—which is very useful for following through on a particular compound or compound type throughout the time period covered. There are 27 volumes for the main work and for each supplement, and if, for example, a particular compound appears in Volume 3 for the main work, it will be in Volume 3 in each of the supplements for which additional information appeared on it during the time period covered by that supplement. Neither the third nor fourth supplement was complete by the middle of 1978. Beginning with Volume 17, the third and fourth supplements are combined.

Within Beilstein, all compounds are classified according to structure, and that classification determines where a compound appears in the main work and supplements. The user who wishes to pursue the literature on classes of related compounds may find it useful to investigate the Beilstein classification system in detail before mapping out a search strategy. There are alternative approaches, especially when information is sought on a specific compound. Even alternative approaches, however, may be facilitated by a rudimentary knowledge of the Beilstein classification scheme. An important aspect of the Beilstein system is that certain kinds of compounds are considered index compounds, while others are "derivatives" of index compounds. For example, an amide might be the derivative of the corresponding carboxylic acid. The first step in systematically locating a compound in Beilstein is usually to determine the index compound. T.C. Owen and R.M.W. Rickett present a lucid but incomplete account of the Beilstein system in Chapter 10 of Bottle's *The Use of Chemical Literature* (Ref. 2.2.) and include a pair of tables that are useful in identifying in which volume a particular compound will be found.

For relatively common organic compounds, a Beilstein entry may be located *via* the *CRC Handbook of Chemistry and Physics* (Ref. 12.4) or some other reference books. The index of this guide identifies such sources under "Beilstein entries". Beilstein has a number of useful formula and name indexes. There is a pair of comprehensive indexes covering the main work and first and second supplements. Users who are unfamiliar with the older German nomenclature generally find the formula index much easier to use than the name index. Since there are usually many isomers under a given formula, knowing in which volume the compound in question will appear is a considerable help. That knowledge comes from an understanding of the classification scheme. There are individual volume indexes for the third and fourth supplements. Some of these indexes have been prepared to cover more than one volume, as for example one index covering both Volumes 9 and 10. A cumulative index to the entire set has recently begun to appear. The parts of this index are for specific volumes—for example there are now formula and name indexes covering the main work and all supplements of volume one. To effectively use this index, it is necessary to know in which volume a compound is going to appear. The nomenclature in these cumulative indexes corresponds to that in the fourth supplement, which is in better accord with IUPAC rules than that used in the earlier parts. The Beilstein entries include information about the preparation, reactions, structure, characterization, and physical properties of organic compounds, and some data are included as well as citations to original publications.

AIDS TO USING BEILSTEIN

14.2   Vestling, Martha M., and Janice T. Liebe. *A Guide to Beilsteins Handbuch der Organischen Chemie*. Boca Raton, FL: J. Huley Associates, 1975. Synchronized slide/audio cassette. 32 slides, 1 audio cassette, 1 eight page booklet.

•   Although the difficult classification scheme is not tackled by this program, it can serve as an introduction to Beilstein. Problems that the user might have with the format, language, and abbreviations in Beilstein, as well as the scope and overall utility of the work, are addressed.

14.3   Runquist, Olaf Allan. *A Programmed Guide to Beilstein's Handbuch*. Minneapolis: Burgess, 1966. 53 pp.

•   This excellent programmed guide is intended to help the student who has had an introduction to organic chemistry learn to use Beilstein by independent study. Strong emphasis is placed on the classification system and arrangement.

14.4   Weissbach, Oskar. *The Beilstein Guide: A Manual for the Use of Beilstein's Handbuch der Organischen Chemie*. Berlin: Springer, 1976. 95 pp.

•   Most of this book is given over to a detailed but terse analysis of the rules for the arrangement of compounds in Beilstein. One of the appendices is a glossary of English and French equivalents of words and expressions frequently used in Beilstein.

GMELIN

14.5   Gmelin, Leopold. *Gmelins Handbuch der Anorganischen Chemie.* 8th ed. Leipzig: Verlag Chemie, 1924-. Multivolume.

• The arrangement of Gmelin is based on the periodic table. However, the arrangement is not obvious. In order to use Gmelin, it is helpful to know which system number has been assigned to each element. A periodic table containing Gmelin system numbers can be found on the front inside cover of many of the volumes of Gmelin. A given inorganic compound will be treated in the volume for the component element with highest Gmelin system number. Each volume, then, is devoted to a specific element (except Volumes 1 and 71 which are devoted to sets of related elements). Within each volume, the earliest parts deal with such topics as the history, occurrence, and exploitation of the element, after which the physical and chemical properties of the element itself are treated, and then its compounds (usually the major portion of the volume). Compounds are arranged according to increasing classification numbers of the component elements. Gmelin is, in effect a comprehensive treatise of inorganic chemistry. The text, which is in German, incorporates a great deal of data, and is augumented with tables, charts, and drawings. The literature is thoroughly covered and well referenced. English language tables of contents and marginal notes help the user who is not familiar with German. Some volumes of Gmelin are much more up to date than others, as different volumes have been issued at different dates. Many volumes are updated by supplements. In addition, there is a supplementary series of Gmelin which is independent of the Gmelin classification scheme, in which unified, authoritative treatments of contemporary research issues are presented. Organometallics are included in the supplementary series. Beginning in 1975, a comprehensive, multivolume formula index to the main and supplement volumes and supplementary series began to appear.

15

Guides to
Techniques

Chemistry is basically a laboratory science. Throughout the history of the subject, there has been consequential work done by calculation, and recently some very important chemical research has been performed by computer. However, all theoretical calculations ultimately have to be supported by the evidence of the experimental laboratory. Two relatively short sections of this chapter deal with computer-based and mathematical methods used by chemists. The remainder is concerned with laboratory techniques, mainly the analysis of the materials to discover their chemical compositions, and the synthetic techniques that are used to prepare chemical substances. The methods by which physical properties of chemical substances are determined are also very important and are perhaps best covered in *Techniques of Chemistry* (Ref. 15.6) as well as in monographs and laboratory manuals too numerous to be included here. The data compilations in Chapters 12 and 13 often include descriptions of measurement techniques or references to sources where such descriptions can be found.

The two "formularies" mentioned in this chapter perhaps do not really belong in this guide, but they have been included because the practical information they contain is often sought (usually in vain) in chemistry libraries. Some, but not all, of the general methodology resources described first in this chapter are at the very basic level of chemical technology. They are included because they can help the beginning student, and because even the experienced researcher sometimes needs to refresh his memory on a basic technique.

There is a vast literature treating the techniques and methods of chemistry. This bibliography makes no attempt to even approach comprehension, but rather is limited to the most significant works along with a sampling of a few other reference materials.

Books on analytical techniques are assigned Library of Congress classification numbers between QD 71 and QD 142. Many, perhaps most, of the books within this classification range include information pertinent to the methodology of chemical analysis. Many publications present techniques of analytical chemistry for specialized situations, such as *Analysis of Ancient Metals* by Earle R. Caley, or detailed applications of specific techniques, such as *Colorimetric Methods of Analysis* by Foster Dee Snell and Cornelia T. Snell. In addition, there are many excellent textbooks, such as *Advanced Analytical Chemistry* by Louis Meites and Henry C. Thomas, which include details of techniques. The appropriate Library of Congress subject heading for analytical chemistry in general is "Chemistry, Analytical." More specific topics may sometimes be found in the general textbooks, or may be sought as books under specific subject headings in the library's catalog.

Books dealing with or including preparative techniques may be found primarily between QD 146 and QD 196 (inorganic chemistry) and between QD 241 and QD 441 (organic chemistry). There are monographs and sets which deal with specific techniques of wide application. An outstanding publication of this type is George A. Olah's multivolume *Friedel-Crafts and Related Reactions*. It is usually possible to locate such books *via* the library's subject catalog under the most specific subject headings, such as "Friedel-Crafts reaction." It is usual library practice to assign only the most specific subject headings that apply to an entire book. Hence, *Birch Reduction of Aromatic Compounds* by A.A. Akhrem, J.G. Reshetova, and Yu. A. Titov, will, in most libraries, be listed under "Birch reduction," but *not* under "Reduction, Chemical."

There are also many books dealing with the preparation or analysis (or both) of specific chemical substances. QD 181 is assigned to books on individual elements or groups of related elements, such as *Analytical Chemistry of Nickel* by Clyde L. Lewis and Welland L. Ott, or *Boron-Nitrogen Compounds* by Kurt Niedenzu and John W. Dawson. The primary literature, mainly journal articles, is, of course, where most analytical and preparative techniques and procedures can be found, described in detail. The indexes, bibliographies, and other reference works described in this guide, such as *Chemical Abstracts* (Ref. 3.1.) and *Current Abstracts of Chemistry* (Ref. 4.8) are especially useful for locating information in the primary literature.

Some of the reference books included in this chapter report analyses or preparations which are not described elsewhere. More often, they compile techniques which have already been published, and they document their sources. However, they are, for the most part, self-sufficient—they contain all of the information that is necessary to carry out the procedure, so it is not usually necessary to refer to the original literature.

Techniques of laboratory safety and of manuscript preparation are the subjects of the two subsequent chapters.

15.1 Chemical Technician Curriculum Project. Writing Team. *Chemical Technology Handbook: Guidebooks for Industrial Chemical Technologists and Technicians.* Edited by Robert L. Peksok, Kenneth Chapman and Wade H. Ponder. Washington: American Chemical Society, 1975. 215 pp.

• Safety precautions and first aid are emphasized. Least square, confidence level, and similar calculations are explained in a chapter on the use and interpretation of data. Other topics include drawings and diagrams, and laboratory tools and equipment.

15.2 *Methoden der Organischen Chemie (Houben-Weyl).* Stuttgart: Georg Thieme, 1952-.

• Woodburn (Ref. 2.20) considers *Houben-Weyl* an outstanding treatise of experimental organic chemistry, and devotes considerable space to a discussion of its structure and use. Synthesis, analysis, and physical methods of investigation are all covered. Each of the volumes except the first four, deals with a specific class of organic compounds. The text is entirely in German. Good coverage of the primary literature is provided.

15.3 *Methodicum Chimicum: A Critical Survey of Proven Methods and Their Application in Chemistry, Natural Science, and Medicine.* Edited by Friedhelm Korte, and others. New York: Academic. 1974-. Multivolume.

• "The Methodicum Chimicum is a short critical description of chemical methods applied in scientific research and practice. It is particularly aimed at chemists as well as scientists working in associated areas including medicine who make use of chemical methods to solve their 'interrelated' problems" (from the Preface of the Series). There is a strong emphasis on biochemistry. The editors stress that the greatest amount of detail is lavished upon the newer methods which are not adequately described in other review sources. However, the older, standard techniques are included in the discussions and literature references. Because many of the authors are from German-speaking countries, there is perhaps a larger proportion of references to the recent German literature than is usually encountered.

The first part of the set deals with analytical chemistry, including separation techniques. A larger portion of the set is devoted to synthesis—both the principles in planning syntheses, and specific synthetic techniques. Entire volumes, or large parts of volumes, are devoted to the preparation of specific classes of compounds, such as transition metal derivatives, nucleic acids, vitamins, etc. Other volumes are related to the formation of specific kinds of bonds—e.g., Volume 5 for C-O bonds, Volume 6 for C-N bonds.

15.4   Parr, Norman L. (ed.) *Laboratory Handbook.* Princeton, NJ: Van Nostrand, 1963. 1523 pp.

• Authoritative chapters have been prepared by experts on such topics as laboratory design, glassblowing, electronic engineering, vacuum technology, distillation, and many other basic techniques. Over two hundred pages are devoted to the work and equipment of specialized laboratories, such as metallurgical, pharmaceutical, and forensic science laboratories. Careful attention to practical details and good illustrations contribute to the value of this handbook.

15.5   Shugar, Gershon J., Ronald A. Shugar, and Lawrence Bauman. *Chemical Technicians' Ready Reference Handbook.* New York: McGraw-Hill, 1973. 463 pp.

• This elementary and well illustrated aid to basic chemistry lab techniques contains such helpful information as "reuniting the mercury column" (of a thermometer), "selection of filter paper", "vacuum pumps: use, care, and maintenance", and "grades of purity of chemicals." Section Two "has been designed to provide a picture or a drawing of the equipment generally found in most chemical laboratories." This section is useful for matching names with common apparatus.

15.6 *Techniques of Chemistry.* Edited by Arnold Weissberger. New York: Wiley-Interscience, 1970-. Multivolume.

• An earlier series, *Technique of Organic Chemistry*, had an established reputation for reliability among students and researchers in the field. The series *Techniques of Inorganic Chemistry* appeared later as a companion. *Techniques of Chemistry* supersedes these two series. As Weissberger points out in the Introduction to the Series, many of the same techniques are used in both organic and inorganic chemistry. The techniques that are described are particularly (but not exclusively) applicable in the research laboratory. Methods and operations are described in detail, and theoretical background is included.

Volume 1, Physical Methods of Chemistry, is divided into five parts: instruments (including automatic recording and control, computers, electronic circuits, etc.); electrochemical methods; optical, spectroscopic, and radioactivity methods; determination of mass, transport, and electrical-magnetic properties; and determination of thermodynamic and surface properties. The purification and physical properties of organic solvents constitute the subject matter for Volume 2, which includes an extensive collection of tables. As indexes to these tables are given by the various properties, such as boiling point, density, and dielectric constant, they are helpful for selecting an appropriate solvent for a specific application. The elucidation of organic structures by physical and chemical methods is presented in Volume 4. Many of these techniques, such as nmr spectroscopy, are subjects of entire books, including some that are discussed elsewhere in this guide. Others, such as the techniques for the assignment of stereochemical configuration, are not so easily located elsewhere.

Techniques for studying kinetics and reaction mechanisms can be found in Volume 6. Other volumes treat such topics as photochromism, electroorganic synthesis, and chromatography.

ANALYSIS

15.7 American Chemical Society. Committee on Analytical Reagents. *Reagent Chemicals: American Chemical Society Specifications.* 5th ed. Washington: 1974. 685 pp.

• "The specifications . . . are intended to serve for reagents to be used in precise analytical work" (page 1). The arrangement is alphabetical by reagent, and for each one, purity requirements, and methods of testing for purity are given.

15.8   American Public Health Association. *Standard Methods for the Examination of Water and Wastewater*. 14th ed. Washington: 1975.

• Chemical, biological, and physical methods of examination are described in this practical, authoritative manual.

15.9   American Society for Testing and Materials, *Book of A.S.T.M. Standards*. Philadelphia: 1916-. Annual.

• This highly authoritative multivolume compilation of standard tests and procedures includes a number which may be of use in the analytical chemistry laboratory. Emphasis is on materials of technological importance. The last volume each year is an index to the entire set.

15.10 Association of Official Analytical Chemists. *Official Methods of Analysis*. 12th ed. Washington: 1975. 1094 pp.

• The emphasis is on agricultural chemicals, but there are also sections dealing with the analysis of food, cosmetics, drugs, and other materials which are not entirely within the domain of agriculture. Changes in the methods are issued in the *Journal of the Association of Official Analytical Chemists*, and are also reprinted as supplements. New editions of the Methods are published every five years. However, every edition since the tenth contains some methods which are not repeated in subsequent editions.

15.11 Busev. A.I., V.G. Tiptsova, and V.M. Ivanov. *Handbook of the Analytical Chemistry of Rare Elements*. Ann Arbor, Michigan: Ann Arbor-Humphrey Science Publishers, 1970. 402 pp.

• Translation of the Russian book which was originally published in Moscow in 1966. Techniques, selected by the authors as the simplest and fastest, are given for 39 of the less common elements.

15.12 *Chemical Analysis of Ecological Materials*. Edited by Stewart E. Allen. New York: Wiley, 1974. 565 pp.

• Techniques for the analysis of soils, vegetation, water, and pollutants are described. For the analysis of ecological materials, sampling techniques are very critical, and careful attention is given to details of collecting and handling samples as well as their analyses. There are also discussions of instrumental and statistical techniques and data processing.

15.13 Clarke, H.T., and B. Haynes. *A Handbook of Organic Analysis, Qualitative and Quantitative.* 5th ed. London: Edward Arnold, 1975.

• In addition to providing details for basic analytical procedures, the authors have included a very extensive table of physical properties of organic compounds (Chapter 6). This table is divided into sections for each class of compound, with the compounds within each section being arranged by increasing boiling or melting point. The book is intended primarily for students.

15.14 *Encyclopedia of Industrial Chemical Analysis.* See Ref. 10.11.

15.15 Erdey, Laszlo. *Gravimetric Analysis.* Oxford: Pergamon, 1963-1965. 3 vol.

• Originally published in Hungarian by Akadémiai Kiadó, Budapest.
Gravimetric methods depend upon the determination of the weight of a precipitate. They constitute the oldest branch of quantitative analysis, being based on the stoichiometric principles which evolved at the beginning of the modern era of chemistry. Even today, many of the standard quantitative techniques, particularly for analyses of technological importance such as determining the composition of ores, are gravimetric. Volume one of *Gravimetric Analysis* gives general principles and techniques; volumes two and three give determinations of the different elements.

15.16 Feigl, Fritz. *Spot Tests in Inorganic Analysis.* 6th ed. Amsterdam: Elsevier, 1972. 669 pp.

• Spot tests are relatively rapid identification techniques, usually dependent upon a color change or some other readily sensed transformation. They are often carried out on specially designed porcelain plates, or on filter paper, employing a drop of a test solution. This book and Feigl's companion volume on organic spot tests (Ref. 15.17) are authoritative and comprehensive.

15.17 Feigl, Fritz. *Spot Tests in Organic Analysis.* 7th ed. Amsterdam: Elsevier, 1966. 772 pp.

● Spot tests are employed for the detection of elements or functional groups in organic compounds, and to a lesser extent, the identification of individual organic compounds. There are some other applications, such as the differentiation of isomers. Feigl's clear and detailed compendium of procedures is the recognized authoritative manual for these important laboratory techniques. A companion volume for inorganic analysis (Ref. 15.16) is listed separately.

15.18 Fitton, A.O., and J. Hill. *Selected Derivatives of Organic Compounds. A Guidebook of Techniques and Reliable Preparations.* London: Chapman and Hall, 1970. 53 pp.

● ". . . primarily designed to enable students to prepare crystalline derivatives as part of the analysis of an unknown substance." (from the Preface).

15.19 Kodama, Kazunobu. *Methods of Quantitative Inorganic Analysis. An Encyclopedia of Gravimetric, Titrimetric and Colorimetric Methods.* New York: Interscience, 1963. 507 pp.

● Step-by-step procedural details are usually not given. However, there are many useful flow charts and tables, in addition to information about reagents, tests, etc., particularly with regard to the determination of specific elements.

15.20 Maddalone, R.F., and S.C. Quinlivan. *Technical Manual for Inorganic Sampling and Analysis.* Springfield, VA: National Technical Information Service, 1977. 334 pp.

● Technical Report No. PB-266842. "The manual presents the state-of-the-art of inorganic sampling and analysis procedures in a standardized format that makes the methodology readily available to professionals in the field" (from the Abstract). The preparation of the manual was sponsored by the Environmental Protection Agency. Analysis for pollutants is stressed.

15.21 Meites, Louis (editor). *Handbook of Analytical Chemistry.* New York: McGraw-Hill 1963.

• This is an extensive compilation of tables relevant for basic techniques of qualitative and quantitative analysis, such as schemes for the classical hydrogen sulfide techniques of determining metal ions, standard procedures for measuring pH, and titrations. Although some techniques are briefly described, the most significant feature of this handbook is its presentation of tables of data that are useful for analyses.

15.22 Scientific Committee on Problems of the Environment (SCOPE). Working Group on Methodology of Determination of Toxic Substances in the Environment. *Environmental Pollutants. Selected Analytical Methods (SCOPE 6).* Ann Arbor, Michigan: Ann Arbor Science, 1975. 277 pp.

• SCOPE, a committee of the International Council of Scientific Unions, is composed of prominent scientists from all over the world. The working group selected what they considered to be good methods for the determination of some of the principal substances of importance in environmental problems. Only some of these methods, however, have been labelled as reference methods; the others "should be regarded as offering guidance in the choice of methods rather than as mandatory procedures" (from the Introduction).

15.23 Siggia, Sidney. *Instrumental Methods of Organic Functional Group Analysis.* New York: Wiley-Interscience, 1972. 428 pp.

• This book is intended by the author to supplement his earlier book, *Quantitative Organic Analysis via Functional Groups,* which covered the more traditional wet (or non-instrumental) techniques. Both books are arranged by functional groups, and procedural details are given.

15.24 *Standard Methods of Chemical Analysis.* 6th ed. Princeton, NJ: Van Nostrand, 1962-1966. 3 vol. in 5.

• Previous editions of this major reference work were usually referred to as "Scott's Standard Methods of Chemical Analysis." In Volume 1, each chapter is devoted to a specific element. Volume 1 also includes several appendices, including one which gives instructions for the preparation of commonly used reagents, standard solutions, and indicators. Volume 2 deals with industrial and non-instrumental methods, and Volume 3 with instrumental methods.

15.25 *Standard Methods of Clinical Chemistry.* New York: Academic, 1953-1972. Irregular.

• The volumes are prepared by the American Association of Clinical Chemists, and present details of time-proven procedures for clinical laboratory analyses.

COMPUTER PROGRAMS

15.26 *Computer Program Abstracts.* Washington. National Aeronautics and Space Administration, Office of Technology Utilization, 1969-. Quarterly.

• Programs developed for or by the National Aeronautics and Space Administration or the Department of Defense are abstracted. There are some of interest in chemistry.

15.27 *Computer Programs for Chemistry.* New York: W.A. Benjamin, 1968-. Irregular.

• Programs applicable to various aspects of chemical research are submitted, and then screened and tested before publication. A much larger number of programs is available through clearinghouses, such as the QCPE (Ref. 15.28).

15.28 Quantum Chemistry Program Exchange. *Catalog and Procedures.* Bloomington, IN: Indiana University, Chemistry Department, Annual.

• The Quantum Chemistry Program Exchange (QCPE) serves as a clearinghouse for digital computer programs for all areas of chemistry, although there is a stated preference for quantum chemistry. The programs listed in the catalog may be ordered from QCPE. As each catalog lists all available programs, it can be considered to supersede all previous annual catalogs. Quarterly supplements and newsletters update the catalog.

DATA HANDLING—MATHEMATICAL TECHNIQUES

15.29 Bauer, Edward L. *A Statistical Manual for Chemists.* 2d ed. New York: Academic, 1971. 193 pp.

• ". . . written for chemists who perform experiments, make measurements, and interpret data" (from the Preface to the first edition). Many formulas, accompanied by worked examples, are presented—for example, the calculation of confidence limits.

15.30 Lark, P.D., B.R. Craven, and R.C.L. Bosworth. *The Handling of Chemical Data.* Oxford: Pergamon, 1968. 379 pp.

- Mathematical techniques drawn from statistics, probability theory, relationships among variables, and other areas are treated in terms of their applications in experimental chemistry. The authors point out in the Preface that they are primarily concerned with elementary techniques, but not entirely limited to them. Tables of statistical functions are appended.

FORMULARIES

15.31 *The Chemical Formulary.* Edited by H. Bennett. New York: Chemical Publishing Co., 1933-. Annual.

- The formulations that are described in this book are mixtures and blends of chemicals for household or industrial use, including cosmetics, detergents, polish, rubber, etc. Much of this information is difficult to locate elsewhere. A cumulative index was issued in 1972, but it does not completely reproduce all the information in the individual indexes. Supplier address information is also included.

15.32 Stark, Norman H. *The Formula Manual.* Cedarburg, WI: Stark Research Corp., 1973. 80 pp.

- "Do-it-yourself" instructions are given for household chemicals such as cleansers, deodorants, and ski wax.

INSTRUMENTATION

15.33 Instrument Society of America. *Standards and Practices for Instrumentation:* 4th ed. Edited by Glenn H. Harvey. Pittsburgh: 1974.

- All of the Society's standards are presented in full. In addition, abstracts of relevant standards from other organizations are provided, and accompanied by a subject index.

15.34 Veillon, Claude, *Handbook of Commercial Scientific Instruments. Volume 1. Atomic Absorption.* New York: Dekker, 1972. 174 pp.

Wendlandt, W.W. *Handbook of Commercial Scientific Instruments. Volume 2. Thermoanalytical Techniques.* New York: Dekker, 1974. 234 pp.

• The information in the volumes of this series should be very useful to scientists contemplating the purchase of new instrumentation in one of the areas covered. Information on costs, accessories, and specifications are provided. There are also photographs and some schematic diagrams. Although most of the information was furnished by the manufacturers, the authors have also included some evaluative comparisons of instruments.

PREPARATIONS

15.35 *Annual Reports in Inorganic and General Syntheses.* New York: Academic, 1972-. Annual.

• Individual authors review the literature but do not provide details of experimental techniques. Each volume has an arrangement based on the periodic table, sometimes with additional articles on special topics.

15.36 Brauer, Georg (editor). *Handbook of Preparative Inorganic Chemistry.* 2d ed. New York: Academic, 1963-1965. 2 vol.

• After an introductory discussion of preparative methods, details are given for the preparation of a large number of compounds and elements. Emphasis is on inorganic laboratory reagents that were not commercially available when the set was published.

15.37 *Compendium of Organic Synthetic Methods.* New York: Wiley—Interscience, 1971-. Irregular.

• "*Compendium of Organic Synthetic Methods* is a systematic listing of functional group transformations designed for use by bench chemists, persons planning syntheses, students attending courses on synthetic chemistry, and teachers of these courses" (from the Preface). It is primarily intended as a time-saver regarding literature searching. Details of the techniques are not given, but for that the reader may follow through in the original literature which is referenced here. The principal index is in the form of a matrix at the beginning of each volume which correlates the classes of compounds to be prepared with the classes from which they are prepared.

15.38 Dub, Michael (editor). *Organometallic Compounds; Methods of Synthesis, Physical Constants and Chemical Reactions.* 2d ed. New York: Springer-Verlag, 1966-1968. 3 vol.

• Volume I: Compounds of the Transition Metals; Volume II: Compounds of Germanium, Tin and Lead; Volume III: Compounds of Arsenic, Antimony and Bismuth. Detailed experimental information is not provided. However, for each compound an indication of how it is synthesized is provided, along with references to the original literature, and some of its properties. An index to the three volumes appeared in 1969, and supplements to each volume were issued in the 1970s.

15.39 Fieser, Louis Frederick, and Mary Fieser. *Reagents for Organic Synthesis.* New York: Wiley, 1967-. Multivolume.

• This major reference work is arranged like an encyclopedia, with each entry providing information about a reagent which has been reported as useful in organic synthesis. Most of the reagents are themselves organic, but inorganic reagents are also included. Suppliers are listed for commercially available reagents; otherwise, a brief indication of the method of preparation is given, along with a literature reference. Concise information and literature references are also provided on the application of reagents. There are author and subject indexes, and a very useful index of reagents by type, so that this reference set can serve as a very convenient point of entry into the literature dealing with a general type of reaction, whether it be a "name" reaction (e.g., Baeyer-Villiger oxidation), a method of introducing a specific functional group (e.g., amination), etc.

15.40 *Inorganic Syntheses.* New York: McGraw-Hill, 1939-.

• Manuscripts describing the preparations of specific inorganic compounds in detail are published on the basis of their clarity of presentation and the authenticity of the technique. The latter is determined, in part, by having the experiment repeated in a different laboratory from the submitting one. This set is highly esteemed by researchers in inorganic chemistry.

15.41 Jacobson, Carl Alfred. *Encyclopedia of Chemical Reactions.* New York: Reinhold, 1946-1959. 8 vol.

• Reactants in chemical reactions are arranged alphabetically by chemical symbol or formula, and for each such reactant, separate entries are provided for its reactions with other substances. The appropriate chemical equations are given, along with a brief description of the reaction conditions, and a bibliographic citation. The literature referred to is mainly 19th and early 20th century; the reactions tend to be simple desk-top reactions with common laboratory reagents. This set is useful for the high school and beginning college student because it brings together in one place examples and details that are scattered in many elementary textbooks.

15.42 *Macromolecular Syntheses.* New York: Wiley, 1963-. Irregular.

• Procedures are given for the preparation of specific polymers which either illustrate useful techniques, or are of general interest.

15.43 Mathieu, Jean, and Jean Weill-Raynal. *Formation of C-C Bonds.* Stuttgart: Georg Thieme, 1973-. Multivolume.

• The formation of carbon-to-carbon bonds comprise the most important classes of synthetic reactions in organic chemistry, for it is through them that the organic skeletal structures can be built. There are great many different reaction types which result in C-C bond formation. This set presents these in outline format, with good graphics, and many examples from the literature. Emphasis is on recent developments.

15.44 *Organic Compounds: Reactions and Methods.* New York: IFI/ Plenum. Irregular.

• This translation from the Russian is similar to *Organic Reactions* (Ref. 15.45), but does not duplicate the topics therein. Literature review sometimes emphasizes Russian publications.

15.45 *Organic Reactions.* New York: Wiley, 1942-. Annual.

• Individual chapters present comprehensive treatments of important classes of organic reactions. Each volume contains a detailed subject index to itself, and cumulated author and broad subject indexes to the whole set. *Organic Reactions* is perhaps the most important source of reviews of the synthetic and other reactions that are used by the organic chemist.

15.46 *Organic Syntheses: An Annual Publication of Satisfactory Methods for the Preparation of Organic Chemicals.* New York: Wiley, 1921-. Annual.

• Details are provided for the syntheses of specific organic compounds, with emphasis in recent years on compounds of general interest, and model compounds. Very high standards are applied. The syntheses which are reported in this set are checked by repetition in independent laboratories. Authors are instructed that their reports must be based on optimal results, and must be clear and complete in detail. All safety hazards must be mentioned.

Organic Syntheses is now indexed and abstracted in *Chemical Abstracts*, although the earliest of its volumes were not. Collective Volumes of *Organic Syntheses* cumulate and update the reports from the previous ten years. For the years for which Collective Volumes are available, they should be used in preference to the individual annual volumes.

The Cumulative Indices to the first five collective volumes were published in 1976. Only in very recent years has *Chemical Abstracts* nomenclature been purposely included in *Organic Syntheses*. However, the Cumulative Indices provide entries for all compounds whose syntheses have been described, both by the *Chemical Abstracts* (Ninth Collective Period) name and the name originally used in *Organic Syntheses*. There are also indexes by reaction type, compound type, formula, solvents, reagents, apparatus, and author, and a general index to the compounds that were mentioned in the text of *Organic Syntheses*, not only those that were the objects of specific preparations. For example, intermediates are listed in the general index.

15.47 *Preparative Inorganic Reactions.* New York: Interscience, 1964-1971. 7 vol.

• The literature dealing with preparative reactions for classes of compounds, or with general preparative topics, is critically evaluated by experts.

15.48 Sandler, Stanley R., and Wolf Karo. *Organic Functional Group Preparations.* New York: Academic, 1968-1972. 3 vol.

• Each chapter treats the preparative methods for a given functional group, and includes details for some specific examples.

15.49 Sandler, Stanley R., and Wolf Karo. *Polymer Syntheses.* New York: Academic, 1974-. Multivolume.

• On the basis of a review of the recent journal and patent literature, the authors present detailed laboratory instructions for the preparation of polymers. Each chapter deals with a particular chemical class of polymer.

15.50 Weygand, Conrad, and A. Martini (editors). *Weygand/Hilgetag Preparative Organic Chemistry.* New York: Wiley, 1972. 1181 pp.

• Translated from German, and prepared as a fourth edition of part of the classic publication of Conrad Weygand, this book is divided into four parts: reactions on the intact carbon skeleton; formation of new carbon-carbon bonds; cleavage of carbon-carbon bonds; and rearrangements. There are appendices on purification of solvents and gases, and preparative organic work with small quantities.

16
Safety Manuals and Guides

In addition to the guides and manuals listed here, there are numerous books that list common chemicals, drugs, and household products along with their toxic properties. In the Library of Congress classification system, these are usually classified at RA 1211.

In *Chemical Abstracts*, in addition to finding entries like "Beryllium, biological studies . . . poisoning by" in the Chemical Substance Index, there are entries in the General Subject Index under topics such as "Toxicology" which refer to articles of a general nature. Beginning in recent years, the keyword indexes have included "safety" as a keyword for all documents which have any information on the safety of chemical reactions or chemical handling operations, and, of course, this means that these same documents will be retrieved by the keyword "safety" from a search on the CA Condensates database. The National Library of Medicine indexes the literature for the toxic effects of drugs and chemicals in its bimonthly *Toxicity Bibliography*. Relevant literature can also be found *via Exerpta Medica*, especially Section 30 (Pharmacology and Toxicology) and Section 35 (Occupational Health and Industrial Medicine). The former places emphasis on drugs, whereas the latter treats a very broad spectrum of occupational health aspects, and includes a subsection on chemical environment. Bio Sciences Information Service publishes *Abstracts on Health Effects of Environmental Pollutants*, which is made up of entries selected from the BIOSIS PREVIEWS database. Medicinals are excluded from coverage. *Engineering Index* gives access to articles on topics such as "Chemical Plants—Accident Prevention." The *CIS Abstracts*, published eight times a year by the International Labour Office in Geneva, includes a section on chemical, oil, fuel, plastics and rubber industries, which reviews safety-related books, manuals, and periodical articles.

16.1 American Chemical Society. Committee on Chemical Safety. *Safety in Academic Chemistry Laboratories.* Washington: 1974. 40 pp.

• The recommendations for safe procedures and facilities are concisely and forcefully presented in this booklet.

16.2 Bretherick, L. *Handbook of Reactive Chemical Hazards. An Index Guide to Published Data.* London: Butterworths, 1975. 976 pp.

• Fire or explosion hazards in handling specific chemicals or in carrying out certain kinds of reactions are concisely described, based on reports that have been published in the primary or secondary literature. Toxic hazards are not covered. Laboratory reagents that are not usually included in the industrially-oriented safety manuals are often included in this work.

16.3 Gray, Charles Horace (ed.) *Laboratory Handbook of Toxic Agents.* 2d ed. Englewood, NJ: Franklin Publishing Co., 1968. 190 pp.

• This convenient little handbook provides safety and first aid information for the major laboratory gases, reagents and solvents, and also includes general laboratory precautions against radiation.

16.4 *Handling Guide for Potentially Hazardous Materials.* Edited by A. David Baskin. Oxford, IN: Richard B. Cross Co., 1975. Loose-leaf.

• Each entry in Section B provides a capsule of risk information, synonyms, precautions, first aid, fire control, and some other related data. The compounds that are handled in commerce are covered. An index by synonym is provided in Section A, at the beginning of the book. Section D provides a medical digest for each of the materials, with separate suggestions for paramedics and physicians. Section E treats pollution control.

16.5 Kaigai Gijutsu Shiryō Kenkyūjo. *Toxic and Hazardous Industrial Chemicals Safety Manual for Handling and Disposal with Toxicity and Hazard Data.* Tokyo: International Technical Information Institute, 1975. 591 pp.

- Safety information is compiled for over 700 commercial compounds. This manual offers the advantage of collecting all of the information for each compound in one place.

16.6 Manufacturing Chemists' Association. *Chemical Safety Data Sheet*. Washington: 1953-. Irregular.

- Each data sheet provides information that is essential for the safe handling of a specific industrial chemical. Revised data sheets are issued as needed. The nature of hazards, handling and storage information, waste disposal, and medical and first aid aspects are discussed in detail. Relevant physical properties are tabulated at the beginning of each data sheet.

16.7 Manufacturing Chemists' Association. *Guide for Safety in the Chemical Laboratory*. New York: Van Nostrand Reinhold, 1972. 505 pp.

- Advice regarding safe building design, equipment selection, sampling techniques, etc., is given. Appendix 3 lists a large number of explosive reactions, with literature references. Appendix 4 contains hazard charts, in which hazardous chemicals are listed alphabetically along with some significant properties. Following the hazard charts are waste disposal procedures for classes of dangerous materials.

16.8 National Fire Protection Association. *Fire Protection Guide on Hazardous Materials*. 6th ed. Boston: 1975. 900 pp.

- Data on most of the chemicals in commercial use are included. There is information pertinent to proper storage and handling that can prevent fires. Information of use to professional fire fighters in emergencies is also included. Tables present the flashpoints of over 8000 trade-named liquids, the fire hazard properties of flammable liquids, gases, and solids, explosion and toxicity data on about 400 hazardous chemicals, and a recommended system for the identification of the fire hazard of materials. A particularly noteworthy section provides information about 3550 mixtures of two or more chemicals that have reacted in a hazardous manner at room or moderately elevated temperatures.

16.9 *Registry of Toxic Effects of Chemical Substances.* Washington: U.S. Government Printing Office. 1975-. Annual.

• The National Institute for Occupational Safety and Health is mandated to issue this list of known toxic substances every year. In 1976, there were 21,729 substances in the list. The editors "include all mined, manufactured, processed, synthesized, and naturally occurring inorganic and organic compounds" (6th ed., p. ix). There are many cross references from synonymous names, but trade names are not usually included. Toxic dosages to humans and some animals, as well as data on the dangers of occupational exposure, are provided. The entries in this Registry are very terse, but there are references to fuller reports, including those in the *Federal Register.*

16.10 Sax, N. Irving. *Dangerous Properties of Industrial Materials.* 4th ed. New York: Van Nostrand Reinhold, 1975. 1258 pp.

• "Sax" is often considered the standard reference source on safety in chemistry. Emphasis, of course, is on industrial chemicals, but in the 4th edition, there are data on nearly 13,000 materials—more than is found in most comparable sources.

The book is divided into two parts—white pages and tan pages. The tan pages are arranged alphabetically by chemical, with hazard data for each one. Since essentially the same information applies to many different materials, the user is often referred from specific materials to appropriate sections of the white pages, where general information on radiation, fire protection, cancer risks, toxicology, and other considerations is given. The coded toxic hazard ratings which are reliable qualitative guides are given for each chemical.

16.11 Steer, Norman V. (ed.) *Handbook of Laboratory Safety*, 2d ed. Cleveland: Chemical Rubber Co., 1971.

• "The purpose . . . is to provide convenient information for hazard recognition and control" (from the Preface). A considerable amount of remedial information is included, such as fire and rescue procedures, and first aid. However, as in most safety handbooks, the book is designed to be consulted *before* the hazardous situation arises. In addition to tables of chemical hazards, information is provided on fire, toxic, radiation, electrical, mechanical, and biological hazards.

17
Style Manuals and Guides for Authors

Authors intending to submit manuscripts for publication in a journal (or elsewhere) should insure beforehand that they meet the writing style requirements of the publishers. For a specific journal, the essential requirements are usually stated in one of the recent issues. The issues which do not carry full details for authors should refer potential contributors to the exact location of such details. Details may vary considerably from one journal to another (even for the same publisher).

Numerous style manuals for the scientific writer have been published. If the Library of Congress classification scheme is used, they can usually be found at T 11. The scientist or technologist can almost inevitably expect writing to be a significant facet of his or her professional work.

Conventions in use of symbols and abbreviations, nomenclature, bibliographic citation format, and graphics often need to be followed. An entire chapter in this guide is devoted to reference sources on nomenclature. Standards organizations have published standards for such conventions, and these standards have been adopted by many publishers. For example, the American National Standards Institute has published the *American National Standard Guidelines for Format and Production of Scientific and Technical Reports* (1974), the *American National Standard for the Preparation of Scientific Papers for Written or Oral Presentation* (1972), the *American National Standard for the Abbreviation of Titles of Periodicals* (1970), and the *American National Standard for Bibliographic References* (1977). Some issues of the *CODATA Bulletin* (published by the International Council of Scientific Unions, Committee on Data for Science and Technology) offer guidance on the presentation of data in the literature.

For example, *CODATA Bulletin* number 13, published in December, 1974, has the title The Presentation of Chemical Kinetics Data in the Primary Literature.

17.1   American Chemical Society. *Handbook for Authors of Papers in the Journals of the American Chemical Society.* Washington: 1967. 125 pp.

•   While some of the instructions are specific for American Chemical Society journals, the good guidance on scientific writing presented in this handbook often has more general applicability. Hints to the typist and notes on grammatical usage are included among the useful appendices.

17.2   Gensler, Walter Joseph. *Writing Guide for Chemists.* New York: McGraw-Hill, 1961. 149 pp.

•   Emphasis is on style, grammar, spelling, and other aspects of writing, particularly those for which the chemist often encounters problems or makes mistakes.

18

Biographies and Directories of People

The most famous men and women in the history of chemistry have entire books written about them. These books can be located by looking up the chemists' names in the library's subject catalog. The lives and works of notable chemists of the past are also described in articles in general encyclopedias, and in the multivolume *Dictionary of Scientific Biography* (New York: Scribner, 1970-76). References to biographical articles in *Chemical Abstracts* may be located *via* the keyword indexes of individual issues under the names of the biographees, or *via* the General Subject Index, under the subject heading "Biography." The *Biography Index* (New York: H.W. Wilson, Co., 1946/47-. Quarterly, cumulated annually) is a good tool for locating biographical information about people in all walks of life, including chemistry.

The address and/or a very brief biographical sketch of a contemporary chemist is often needed. A recent article published by the person will usually be accompanied by his or her address. Many secondary services, including *Chemical Abstracts* incorporate the work addresses of authors in their entries. Because this type of information is so important, two excellent general science sources (numbers 18.3 and 18.9) for current professional addresses are included in this chapter. Otherwise, entries are limited to works which are devoted to chemistry and chemical engineering.

18.1 American Chemical Society. *College Chemistry Faculties*. 4th ed. Edited by Bonnie R. Blaser and Jeanann M. Dellantonio. Washington: 1977. 189 pp.

• Chemistry, chemical engineering, and biochemistry faculties of two and four year colleges in the U.S., Canada, and Mexico are listed. Arrangement is by state or province, with indexes by institution and by individual faculty member.

18.2   *American Chemists and Chemical Engineers.* Edited by Wyndham D. Miles. Washington: American Chemical Society, 1976.

 • Short articles tell about approximately five hundred American chemists and engineers from those who lived in Colonial times to those who died recently.

18.3   *American Men and Women of Science.* 13th ed. New York: Bowker, 1976. 7 vol.

 • This is the most important directory of contemporary American scientists. The brief entry on each scientist contains essential biographical information—date and place of birth, education, professional experience, society memberships, research interests, present professional position. Nearly 110,000 people are listed. A one-volume directory devoted exclusively to chemistry has been "cut" from the entire work and is available separately. *American Men and Women of Science* depends on voluntary information submitted by the biographees. There are discipline and geographic indexes.

18.4   Campbell, William Alec, and N.N. Greenwood. *Contemporary British Chemists.* London: Taylor and Francis, 1971. 286 pp.

 • One to two-page biographies, accompanied by portraits, are given for 140 prominent chemists in British academia and industry.

18.5   *Chemical Engineering Faculties.* New York: American Institute of Chemical Engineers, Chemical Engineering Education Projects Committee, 1951-. Irregular.

 • Information, including lists of faculty members, is given for departments in the U.S., and for a selection of foreign departments.

18.6   *Corporate Diagrams and Administrative Personnel of the Chemical Industry.* Princeton, NJ: Chemical Economic Services, 1958-. Biennial.

- The top and middle management personnel are listed for U.S. companies where chemistry or chemical technology are important to the manufacturing process. Organization charts are provided for some companies.

18.7 Farber, Eduard. (ed.) *Great Chemists*. New York: Interscience, 1961. 1642 pp.

- Lengthy biographical essays, accompanied by portraits, are presented for over 100 outstanding chemists of the past.

18.8 Farber, Eduard. *Nobel Prize Winners in Chemistry, 1901-1961*. Rev. ed. London: Abelard-Schuman, 1963. 341 pp.

- Each entry consists of a small portrait, a biographical sketch, and (usually) a very brief description of the prize-winning work.

18.9 Institute for Scientific Information. *I.S.I.'s Who is Publishing in Science*. Philadelphia: 1971-. Annual.

- Authors whose papers were listed in any section of the previous year's *Current Contents* (Ref. 4.9) or *Science Citation Index* are listed here with their work addresses. Because of the extensive coverage of the source publications, this is a very comprehensive address directory.

18.10 *International Chemistry Directory*. New York: W.A. Benjamin, 1969/1970. 1111 pp.

- It is unfortunate that this directory, which lists academic faculties from all over the world, has not been kept up to date.

18.11 Smith, Henry Monmouth. *Torchbearers of Chemistry: Portraits and Brief Biographies of Scientists Who Have Contributed to the Making of Modern Chemistry*. New York: Academic, 1949. 270 pp.

- The notable collection of portraits upon which this book is based are assembled in the chemistry department of the Massachusetts Institute of Technology. Over 200 of the most renowned men and women in the history of chemistry are represented.

19

Product, Service, and Company Directories

Directories have already begun to get outdated by the time they are put on the shelves of the library. It should not be surprising, then, that most of the publications listed in this chapter are published on a continuing basis, at annual or other intervals. Prices of chemicals are particularly subject to change, so that few product directories include prices. Current catalogs of suppliers should be consulted if the availability and cost of a chemical is needed. A number of firms, such as Eastman Organic Chemicals, Aldrich, and the Alfa Division of Ventron Corporation, supply small quantities suitable for research and educational needs. The catalogs from such firms often list thousands of compounds, and they sometimes include melting or boiling points as well as price and purity information. The catalogs of individual firms are not listed in this guide, but directories compiled from several firms are included here in this chapter.

The chapter also includes directories that may be used to locate companies, large and small, that provide chemical products or services. One very important government agency catalog (Ref. 19.18) is included as well.

19.1 American Chemical Society. *Directory of Graduate Research.* See No. 4.2.

19.2 American Council of Independent Laboratories. *Directory. A Guide to the Leading Independent Testing, Research and Inspection Laboratories of America.* Washington: 1937-. Biennial.

- A useful subject index of services is part of the directory.

19.3   American Society for Testing and Materials. *Directory of Test-in Laboratories, Commercial-Institutional.* Philadelphia: 1954-. Biennial.

- "This directory gives the locations of testing laboratories equipped and prepared to undertake testing on a fee basis" (from the Introduction). Laboratories are listed in alphabetical order, along with an indication of the types of tests they perform. There are subject and geographic indexes.

19.4   *Chem BUY Direct: International Chemical Buyers Directory.* Edited by Friedrich W. Derz. Berlin: de Gruyter, 1974-. Multi-volume.

- This major international guide to commercial chemical products is issued in several interrelated parts. Rare chemicals for research, dyes, isotopically labelled compounds, etc. are included among the over 300,000 names listed in the Chem PRODUCT Index. Each name is accompanied by a registry number (the Chemical Abstracts Registry number is used if one has been assigned).

  To find the suppliers of the compounds, the registry number must be located in the Chem SUPPLIERS Directory, to get the code of the supplying firm, and that code in turn must be located in the Chem ADDRESS Book for the full name and address of the supplier. The complex arrangement is presumably necessitated by the extensive coverage of this source (approximately 23,000 firms) and the fact that many chemicals will be supplied by more than one company.

19.5   *Chem Sources—Europe.* Flemington, NJ: Directories Publishing Co., 1973-. Annual.

- Format and arrangement are similar to *Chem Sources—U.S.A.* (Ref. 19.6).

19.6   *Chem Sources—U.S.A.* Flemington, NJ: Directories Publishing Co., 1958-. Annual.

- Over 75,000 organic and inorganic chemicals are listed with codes for the companies which make them. Full addresses and telephone numbers of the companies are given in a section at the back of the book. There is also a section that lists companies according to specific applications.

19.7 *Chemical Guide to Europe*. Park Ridge, NJ: Noyes Data Corp., 1961-.

19.8 *Chemical Guide to the United States*. Park Ridge, NJ: Noyes Data Corp., 1962-.

• The addresses, products, and other facts pertaining to chemical firms are listed.

19.9 *Chemical Industry Directory and Who's Who*. London: Chemical Age, Benn Brothers, 1923-. Annual.

• This is a good source of information about people and firms in the British chemical industry.

19.10 *Chemical Marketing Reporter*. New York: Schnell Publishing Co., 1871-. Weekly.

• Until 1972, the title was *Oil, Paint and Drug Reporter*.
This is a good source for current prices of industrial chemicals.

19.11 *Chemical Week*. New York, McGraw-Hill, 1914-. Weekly.

• The annual Buyers' Guide Issue of this periodical includes chemicals, raw materials, specialties, packaging, shipping, and bulk containers.

19.12 *Guide to Scientific Instruments*. (Guide issue of *Science*.) Washington: American Association for the Advancement of Science. Annual.

• This is an excellent source for locating the manufacturers of instruments (and instrument parts) that are needed for teaching and research laboratories.

19.13 *Industrial Research Laboratories of the United States*. New York: Bowker, 1920-. Irregular.

• Many of the industrial labs in this directory employ chemists and perform research in chemistry or its applications. In fact, it is one of the best sources to go to for information about laboratories that are not academic or governmental. The arrangement is alphabetical by company name, with divisional facilities listed after the headquarters. Addresses of the laboratories, names of the officers of the company and other key people, and a short paragraph describing the activities are included in each entry. Subject, geographic, and personnel indexes are included.

19.14 *International Chromatography Guide.* (Directory issue of the *Journal of Chromatographic Science.*) Niles, IL: Journal of Chromatographic Science. Annual.

• The annual directory of manufacturers and suppliers of instruments, accessories, supplies, and services used in all types of chromatography is usually appended to the February issue of the journal.

19.15 *OPD Chemical Buyers Directory.* New York: Schnell Publishing Co. 1973/74-. Annual.

• With antecedents dating back to the early part of this century, this annual is received as part of a subscription to *Chemical Marketing Reporter* (Ref. 19.10), or it may be purchased separately. In the 1977/78 issue, over 10,000 commercial chemicals and related products were listed with suppliers.

19.16 *Research Centers Directory.* Edited by Archie M. Palmer. 5th ed. Detroit: Gale, 1975. 1039 pp.

• This is the most important directory of U.S. and Canadian university-related and other non-profit research centers. It is updated periodically by *New Research Centers*, by the same publisher.

19.17 Rubber and Plastics Research Association of Great Britain. *New Trade Names in the Rubber and Plastics Industries.* Shawbury, England: 1962-. Annual.

• Each volume lists new names that have come to the attention of the publisher, along with the manufacturer. Coverage is international.

19.18 U.S. National Bureau of Standards. *Catalog of NBS Standard Reference Materials.* 1975-76 Edition. Washington: U.S. Government Priting Office, 1975. 87 pp. (NBS Special Publication 260).

- The U.S. National Bureau of Standards supplies Standard Reference Materials, which the bureau describes as "well-characterized and certified materials, produced in quantity to help develop reference methods of analysis or test . . . and/or to calibrate a measurement system." (From the U.S. National Bureau of Standards, Monograph 148. *The Role of Standard Reference Materials in Measurement Systems.* (1975) p. 4.) The *Catalog of NBS Standard Reference Materials* is issued at irregular intervals, with the latest edition superseding the previous ones.

19.19 *Worldwide Petrochemical Directory.* Tulsa: The Petroleum Publishing Co., 1969-. Annual.

- This directory contains information about companies engaged in the manufacture of chemicals from petroleum, and includes addresses, key personnel, and plants that are operating, under construction, or planned.

19.20 Wyatt, Alan G. *Chemicals 77: A Directory Giving Information on Chemicals and Allied Products Marketed by Member Firms of the Chemical Industries Association.* London: Chemical Industries Association, 1976. 211 pp.

- This directory is for the chemicals manufactured and/or distributed in the United Kingdom. Trade and proprietary names are included.

20
Non-Print Media

The transfer of chemical information may, under certain circumstances, take place *via* non-print media, such as audio-visual software: films, audio tapes, video tapes, etc. The visual media are particularly important in chemical education. Slides, film strips, video tapes, and motion picture films can be used to demonstrate molecular models and to teach fundamental principles that require graphics. Molecular dynamics may be illustrated by animation or the filming of physical models. The audio-visual presentation of laboratory procedures is not only useful for academic curricula but also for training technicians in government and industrial laboratories. Several examples of how audio-visuals contribute to efforts in chemical education are described in the book *Educational Technology in the Teaching of Chemistry* (Oxford: International Union of Pure and Applied Chemistry, 1975) edited by C.N.R. Rao.

There are few directories or guides for non-print media exclusively in chemistry. Hence, most of the resources listed in this chapter are general in their subject scope, but do include chemistry. Audio-visual aids in using reference works are treated with the specific works themselves—see for example Ref. 13.1.4, *Using the Sadtler Spectra*.

In addition to the sources listed here, centers such as the National Audiovisual Center in Washington and the Massachusetts Institute of Technology publish catalogs of their media holdings. Many of the items in these catalogs cannot be found in other sources.

20.1 *AAAS Science Film Catalog.* Edited by Ann Seltz-Petrash and Kathryn Wolff. Washington: American Association for the Advancement of Science, 1975. 398 pp.

   • This is an important catalog of films in all the sciences, including several hundred films on chemistry and chemical technology at the junior high through adult level, and a small number at the elementary school level. Annotations taken from distributor catalogs or promotional material accompany the entries.

20.2   *Index to Instructional Media Catalogs.* New York: Bowker, 1974. 272 pp.

• The producers of a particular type of audio-visual medium in a subject area such as chemistry can be identified through this easy-to-use index. About 50 separate catalogs are listed under "chemistry", classified according to media type and grade level (from pre-school through adult education).

20.3   National Information Center for Educational Media. *Index to Educational Audio Tapes.* 2d ed., 1972; *Index to Educational Videotapes.* 3rd ed., 1974; *Index to 16 mm Educational Films.* 5th ed., 1975; *Index to 35 mm Educational Films.* 4th ed., 1973; *Index to Producers and Distributors.* 2d ed., 1973. Los Angeles: National Information Center for Educational Media, University of Southern California.

• The NICEM (National Information Center for Educational Media) indexes are based on a computerized information bank at the University of Southern California. Coverage of non-book media is nearly comprehensive, and embraces all subject areas, including chemistry. Entries include brief descriptive annotations. The NICEM databank has also been made available for online searching.

20.4   The Royal Institute of Chemistry. *Index of Chemistry Films.* London: 1970. 349 pp.

• Subtitle: A comprehensive list of films, filmstrips and film loops on chemistry and related topics.
   Brief annotations accompany the entries. A few of the films listed are not in English.

21
Monographs, Textbooks, and Treatises

The many pieces that make up contemporary science are usually first published in journal articles, where they have a chance to be critically reviewed by the scientific community, ultimately to be either rejected or incorporated (perhaps with modification) into the scientific paradigm. The state of the art in various areas of chemistry is recorded in review publications—including review articles as well as books (called monographs) devoted to specific topics. Textbooks are specifically designed for teaching a subject, and hence differ from monographs, which aim to inform the reader of the status of the subject. The distinction, however, is not always easy to make, especially for advanced level books. It is, of course, out of the question to provide a list of monographs and textbooks in this guide. Some of the sources listed in previous chapters, especially Chapter 8, are recommended for such listings.

The "grand" picture of a subject is provided by the comprehensive treatise—usually a multivolume work which exhaustively covers (and references) the primary and secondary literature of a field, and presents a critical evaluation and summary. Decades ago, treatises were attempted for the entire subject of chemistry. No such foolhardy attempt has been made recently.

However, there are some excellent modern treatises on relatively broad areas of chemistry. In a sense, the two works treated in Chapter 14, Beilstein and Gmelin, are treatises, but they are treated separately because of their outstanding significance as reference sources. The most important contemporary treatises are listed below. Most of them have open-ended dates—they are still coming out piece-by-piece, and in some cases, no one can predict when the series will end. The reader is reminded that a good treatise is by necessity also a good literature guide, and its systematic subject-related arrangement often provides a useful alternative to the alphabetical approach to the literature through indexes.

21.1   *The Chemistry of Heterocyclic Compounds; A Series of Monographs.* New York: Interscience, 1950-. Multivolume.

- Heterocyclic compounds are organic ring compounds that contain one or more non-carbon atoms as part of the ring skeleton. They constitute a numerically large class, and include a vast number of compounds of practical importance. Each volume in the series is devoted to a specific class of heterocycles, based on structure. Ring systems are drawn on the book spines to help the user select the appropriate volume.

21.2   *Comprehensive Chemical Kinetics.* Edited by C.H. Bamford and C.H.M. Tipper. Amsterdam: Elsevier, 1969-. Multivolume.

- This is a massive set, as well it should be since kinetics studies are so important in understanding (and predicting) chemical reactions. Each volume in the set deals with one topic. The first three volumes treat practice and theory. The remainder of the set has volumes devoted to various kinds of reactions. Organic, inorganic, and polymerization reactions are all within the scope of the set.

21.3   *Comprehensive Inorganic Chemistry.* Edited by J.C. Bailar, Jr., H.J. Emeleus, Sir Ronald Nyholm, and A.F. Trotman-Dickenson. Oxford: Pergamon, 1973. 5 vol.

- The individual chapters, arranged according to a carefully planned system, are each authored by internationally recognized experts. This is one of the foremost authoritative sources in inorganic chemistry.

21.4   *Elsevier's Encyclopaedia of Organic Chemistry.* Edited by E. Josephy and F. Radt. New York: Elsevier, 1940-1956. Multivolume.

- The initial concept of this treatise was to cover all of traditional organic chemistry in four series: aliphatic compounds, carboisocyclic non-condensed compounds, carboisocyclic condensed compounds, and heterocyclic compounds. However, only the series on carboisocyclic condensed compounds was actually produced.

The arrangement is intended to bring structurally related compounds together, with priority given to the skeletal ring structure. Within the areas included, the editors have attempted to cover the literature comprehensively and critically, and to eliminate erroneous data. Each volume has formula and subject indexes, and the latter include functional group entries as well as entries for compound names. Some additional supplements have been published since 1956 by Springer.

21.5   Kolthoff, Izaak Maurits, and Philip J. Elving (Editors). *Treatise on Analytical Chemistry*. New York: Interscience Encyclopedia, 1959-. Multivolume.

• The treatise is divided into three parts: theory and methods; the elements and inorganic and organic compounds; and analytical chemistry in industry. The principles upon which the various analytical techniques are based are given ample coverage. At the same time, the editors aim for the treatise to "be a guide to the advanced and experienced chemist . . . in the solution of his problems in analytical chemistry, whether of a routine or of a research character" (from the Preface to the Treatise).

21.6   Mellor, Joseph William. *A Comprehensive Treatise on Inorganic and Theoretical Chemistry*. London: Longmans, Green, 1927-1937. 16 vol.

• The literature of the first part of this century in inorganic chemistry, much of which is still important, is very comprehensively covered. Beginning in 1956, supplements to some of the volumes have been issued (some of them published by Wiley in New York). For example, in 1971, a 1467 page supplement on phosphorus chemistry was published.

21.7   *Organic Reaction Mechanisms*. London: Interscience, 1965-. Annual.

• This is a series of annual surveys rather than a treatise. However, it is included here because the important topic of reaction mechanisms is an elusive one to search for in the literature.

21.8   Partington, James Riddick. *An Advanced Treatise on Physical Chemistry*. London: Longmans, Green, 1949-1962. 5 vol.

• Most of the references are to the first half of the twentieth century. Vol. 1: Fundamental Principles: The Properties of Gases. Vol. 2: The Properties of Liquids. Vol. 3: The Properties of Solids. Vol. 4: Physico-Chemical Optics. Vol. 5: Molecular Spectra and Structure; Dielectrics and Dipole Moments. As befits an advanced treatise in physical chemistry, much of the treatment is mathematical. Most of the material is "classical" physical chemistry, but the period covered was that when quantum chemistry and statistical mechanics were being developed, and contributions from these areas are not ignored.

21.9   Partington, James Riddick. *A History of Chemistry*. London: Macmillan, 1962-1970. 4 vol.

• This is a scholarly approach, copiously footnoted, to the history of chemistry from the earliest times to the beginning of the twentieth century.

21.10 Pascal, Paul Victor Henri (editor). *Nouveau Traité de Chimie Minérale*. Paris: Masson, 1956-. Multivolume.

• The traditional topics of inorganic chemistry, such as the geochemical distribution of the elements, the mineral resources, and the structures and chemical reactions of the elements and their compounds are treated in detail. Arrangement is based on the periodic table.

21.11 *Physical Chemistry, an Advanced Treatise*. Edited by Henry Eyring, Douglas Hendersen, and Wilhelm Jost. New York: Academic, 1967-. Multivolume.

• "The purpose of this treatise is to present a comprehensive treatment of physical chemistry for advanced students and investigators in a reasonably small number of volumes" (from the Forward).

21.12 *Rodd's Chemistry of Carbon Compounds; A Modern Comprehensive Treatise*. 2d ed., edited by S. Coffey. Amsterdam: Elsevier, 1964-. Multivolume.

- Part I: Introduction and Aliphatic Compounds. Part II: Alicyclic Compounds. Part III: Aromatic Compounds. Part IV: Heterocyclic Compounds. A systematic arrangement prevails within each part. Specific chemical compound names are highlighted in bold letters making them easy to spot. Each volume has its own index, and an index to the entire set is planned. The text is terse, but there is generous provision of references to the original literature.

21.13 Wilson, Cecil L., and David W. Wilson (editors). *Comprehensive Analytical Chemistry*. Amsterdam: Elsevier 1959-. Multivolume.

- The editors remind us that "the good analytical chemist will make use of any tool that comes to hand" (General Introduction), and hence a truly comprehensive treatise in analytical chemistry cannot be written. The editors have decided to concentrate on those topics which are commonly called upon in the conduct of chemical analysis. The articles are clearly written and well illustrated, and in many cases sufficient details are given to provide laboratory guidance. This is a practical treatise, of use where analytical chemistry is practiced.

22

Chemical Information Search Strategy

There can be no "search strategy" that can be applied like a computer algorithm to find information in the chemical literature. There is some degree of uniqueness to every quest. For this reason, no attempt to outline a strategy will be made here. Rather, a selection of fictitious "cases" will be described.

To adequately describe a case, the intellectual history of the information seeker, a detailed description of the collections and services available, and other factors would have to be described at length. In fact, each "case" could almost be a short novel, although lacking the features that usually attract readers to novels. Nothing of the sort is attempted here. Rather, each case is sketched in briefly, with—hopefully—just enough information for the reader to "get the picture."

Eleven cases are described—not nearly enough, really—but they have been chosen to illustrate the use of some of the most important reference tools in chemistry. Specific illustrations are given where one or another resource either succeeds or fails in producing the needed information. Had alternative illustrations been chosen, the outcomes with the same resources might have been different. In general, the information-seeking patterns described are those which the author would expect to be reasonably successful (except when he purposely chose to describe a poor strategy to illustrate a problem). Although alternative paths are given on occasion, no attempt is made to exhaustively describe all alternative paths (or resources) that will lead to the successful conclusion of a case.

Four very important general reference works, which are among the resources consulted in the cases, are listed below:

22.1 *Applied Science and Technology Index.* New York: H.W. Wilson, 1913—. Monthly.

22.2McGraw-Hill Encyclopedia of Science and Technology. New York: McGraw-Hill, 1977. 15 Vol.

22.3Readers' Guide to Periodical Literature. An Author and Subject Index. New York: H.W. Wilson, 1900/01—. Monthly.

22.4Science Citation Index. Philadelphia: Institute for Scientific Information, 1961—. Quarterly.

CASE I

David, an honors undergraduate majoring in chemistry at a large university, is pursuing an independent studies project in which he has to prepare o-nitrobenzaldehyde

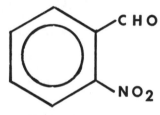

Figure A

(compound A) as part of his research. Although David is able to ascertain from *Chem Sources—U.S.A.* (Ref. 19.6) that the compound is commercially available, his advisor wants him to prepare it himself for the experience he can gain.

David first looks into the Cumulative Indices to *Organic Syntheses* (Ref. 15.46). There he finds two references, III, 641; V, 825, *via* either the formula index (under C_7H_5NO) or the name index. He notes the *Chemical Abstracts* 9th Collective Index name and registry number which are also given in the name index: Benzaldehyde, 2-nitro [552-89-6]. The two references correspond to Volumes III and V of the Collective Volumes of *Organic Syntheses*. David chooses to check the procedure in Volume V first, suspecting that it might supersede the earlier one. However, when he reads the preparation in Volume V, he notes that the author, A. Kalir, reports his preparation as an alternate route to one offered in Volume III, and that no criticism of the earlier procedure is given. David might profit from examining the alternative synthesis as well.

If David wants to find out if any more recent information on this compound has been published, he has several alternative approaches. Since the short article by Kalir is very specifically about the preparation of the compound of interest, a citation search in *Science Citation Index* (Ref. 22.4) has appeal as a rapid way to scan the more recent literature, with little likelihood of totally false leads due to authors citing the paper for other aspects that the student is not interested in. Thus, he might look up A. Kalir in the Citation Index of *Science Citation Index* (for 1973, the year of Collective Volume V of *Organic Syntheses*, and all subsequent years). In doing this, David should recall that the Collective Volumes are cumulations of the annual *Organic Syntheses* volumes of the previous ten years. Kalir had first reported the preparation in the annual volume 46 (1966) on page 81. David thus considers both of these papers as starting points in his citation search. This approach turns up only one article published after 1973. Under "Kalir A" in the 1977 Citation Index, there is the entry

66 ORG SYNTH 46 81

for the cited paper. Under that entry, there is a reference to a paper by S.W. Balasubrahmanyam in the *Journal of Organic Chemistry*—i.e., Balasubrahmanyam cited Kalir's paper because he used Kalir's method to prepare o-nitrobenzaldehyde, but there is actually very little in Balasubrahmanyam's paper about this compound *per se*, and the paper would probably be of very limited interest to David for his particular need at this time.

Science Citation Index does not provide a comprehensive approach to this problem for two reasons. First, the Kalir paper would have to have been cited in one of the approximately 2,500 major scientific periodicals included in SCI's coverage. Non-journal publications, and many of the minor journals are left out, although coverage has been greatly extended recently. Second, and more important in this case, o-nitrobenzaldehyde is an "old" compound—that is, it has been known for a long time, so it is not unlikely that authors who write about it will fail to note Kalir's report on its preparation.

Chemical Abstracts offers a much more comprehensive approach. Since the student is interested only in publications that have appeared since 1973, he can use the Ninth Collective Chemical Substance Index (Ref. 3.1.2), and the individual volume indexes for the tenth collective period. The name "Benzaldehyde, 2-nitro" which he found in the Cumulative Indices to *Organic Syntheses*, will unquestionably be

applicable within this time period. In the collective index, David finds
several hundred references to his compound. He must look at the
index modifications to choose those he is particularly interested in
reading. For instance, one of the modifications reads "sepn. of, from
isomers by gas chromatog.," a subject that might be of some interest to
someone who is trying to prepare the compound. Note, however, that
because of the comprehensive coverage of *Chemical Abstracts* many
of the publications located there will be difficult to obtain, or in unfamiliar languages.

If David wishes to obtain the most current information that *Chemical Abstracts* can offer him, he will have to also consult the Keyword
Indexes of the recent individual issues for which volume indexes have
not yet appeared. He can cut his time in half for this tedious search by
looking only at the odd-numbered issues—i.e., those which include
organic chemistry. The Keyword Index search will be relatively less
satisfactory (in comparison with the Chemical Substance Index search)
because of the lack of vocabulary control and the relative shallowness
of indexing in the Keyword Indexes.

Alternatively, David might use the *Chemical Abstracts* registry
number 552-89-6, which he obtained from the Cumulative Indices to
Organic Syntheses, to do a computer search on Lockheed's files 3 and
4 (see Ref. 5.4). In this way he can rather rapidly search the literature
on his compound, and bypass the look-ups in the Chemical Substance
and Keyword Indexes. If he wishes to limit his computer search to
only certain aspects of the compound, say its preparation, purification, and analysis, a search analyst who is familiar with the data base
will know the best procedure for doing that. However, since most academic libraries cannot offer free computer searches, David will most
likely have to pay at least part of the costs out of his own pocket, if he
chooses the computer search option.

CASE II

The upper division chemistry undergraduates at the university where
Lola is studying can receive credit towards their degree by preparing a
lengthy report that reviews the important literature on a topic that one
of the professors is interested in. Lola's advisor wants her to report on
the most significant applications and mechanistic studies of the aldol
condensation. He informs her that a review of the subject had

appeared around 1967 or 1968 in *Organic Reactions* (Ref. 15.45), and he wants her to report on the developments that have occurred since that review appeared.

Lola looks into the Chapter and Topic Index of the latest volume of *Organic Reactions* on the library shelves, and finds an article in Volume 16 (1968), "The Aldol Condensation" by Arnold T. Nielsen and William J. Houlihan. After reading the article, to get a good background on the subject, she begins her library search for more current information. Because she does not have very extensive knowledge of the subject, and she is to select important information rather than prepare a comprehensive report, she is particularly anxious to find additional review material.

Fieser and Fieser's *Reagents for Organic Synthesis* (Ref. 15.39) is a particularly valuable reference work, at least for the "synthesis" aspect of the subject, because the authors are international authorities who can be counted on to highlight the most important aspects of a topic, and because the few references that are provided are highly selective. Lola thus consults the more recent volumes of this source and gets a good overview of some important recent developments as well as a few helpful references to original articles. Lola goes to the annual volumes of *Organic Reaction Mechanisms* (Ref. 21.7) for references to some of the more theoretical aspects of the subject.

The Index Guide of *Chemical Abstracts* indicates that "aldol condensation" is itself an entry in the General Subject Index (Ref. 3.1.3), and also suggests additional terms for her to pursue. Lola then looks up "aldol condensation" and the other terms in the General Subject Index, and notes in particular those abstract numbers labelled with "R" indicating "review."

Lola also looks in *Science Citation Index* (Ref. 22.4) to see who has cited Nielsen's review article. In particular, a more up-to-date review article can be expected to cite the earlier major reviews, and hence should be listed under Nielsen's article. If the citing article is a review, it will be identified with "R," as in the entry

HO TL CHEM REV R 75 1 75

which is listed under Nielsen's article. Thus, Lola identifies a review article by T.L. Ho, published in Volume 75 of *Chemical Reviews*, which cites Nielsen's article. Unfortunately, this particular article is a review of a different subject, and has very little to offer Lola for her project. In fact, from both the *Chemical Abstracts* and *Science*

Citation Index searches, Lola concludes that, while there have been some English language review articles that touch on one or another aspect of the aldol condensation, there have been none since that by Nielsen and Houlihan which are comprehensive. Nonetheless, she is able to locate non-review articles of interest through these two sources.

CASE III

Charlie, the chemistry librarian in a small industrial firm, has been assigned the task of composing a comprehensive bibliography on aminoindazoles. Figure B illustrates the indazole ring. Although Charlie has a bachelor's degree in chemistry, he is not very familiar with the

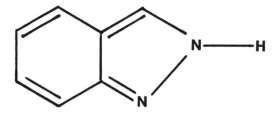

Figure B. Indazole

chemistry of this specialized class of compounds, so the first thing he does is to read the chapter on indazoles in the treatise *The Chemistry of Heterocyclic Compounds* (Ref. 21.1). There is, in fact, a section of that chapter on aminoindazoles, and Charlie is able to obtain several references from that source. From the information in the treatise, Charlie realizes that he is going to have to cover literature as far back as the nineteenth century. Hence, he decides to pursue his search in Beilstein (Ref. 14.1). One of the quickest ways to locate a compound in Beilstein is to find the appropriate Beilstein reference in a standard reference source, such as the *CRC Handbook of Chemistry and Physics* (Ref. 12.4). However, Charlie is unable to locate any aminoindazoles there. He is able to get three Beilstein references from the CRC *Atlas of Spectral Data and Physical Constants for Organic Compounds* (Ref. 12.6):

4-aminoindazole (B25², 308), 6-aminoindazole (B25², 308), and 7-aminoindazole (B25, 318).

B25, 318 refers to page 318 in volume 25 of the main work; B25², 308 refers to page 308 in volume 25 of the second supplement. Because of Beilstein's systematic arrangement, any other aminoindazole will be easily located nearby. Volume 25 has not yet been published for the third or fourth supplements, so Charlie cannot cover the literature after 1929 with Beilstein.

Charlie could have approached Beilstein directly by making use of the Beilstein classification scheme based on molecular structure. Indazoles are heterocyclic compounds with two skeletal nitrogens. According to Table 10.1 in Bottle's *The Use of Chemical Literature* (Ref. 2.2), compounds meeting these specifications are to be found in volume 25. An indazole with one amino substituent will have the formula $C_7H_7N_3$. The Beilstein references can thus be readily found *via* the combined formula index to the main work and the first two supplements. If Charlie must include indazoles with more than one amino group, he should also look under the appropriate formulas, such as $C_7H_8N_4$, for compounds with two amino groups.

He proceeds to *Chemical Abstracts* for the post-1929 literature. Beginning with the Third Decennial Subject Index (1927-1936) and proceeding forward in time, he is able to find several references for the compounds. Up to the Sixth Collective Index, the appropriate entries are of the form

Indazole
———, 5-amino-

Beginning with the Sixth Collective Index, there is a slight change in the name from "Indazole" to "1*H*-Indazole," to distinguish it from isomers such as "2*H*-Indazole." "2*H*-Indazole" had previously gone under the name "Isoindazole." A somewhat more serious nomenclature change occurred beginning with the Ninth Collective Period (i.e., beginning in 1972) for from 1972 on, the amino group in this compound must be indicated as a suffix rather than a prefix. Thus, "3-amino-1*H*-indazole" must now be named "1*H*-indazole-3-amine." Should Charlie be puzzled by the "disappearance" of his compound from *Chemical Abstracts* in 1972, he might consult the *Chemical Substance Name Selection Manual* (Ref. 3.4) to find out how amines are currently named in *Chemical Abstracts*.

For assurance of comprehensive coverage, Charlie also searches the German publication *Chemisches Zentralblatt* (Ref. 4.7) from 1930 to 1969, when it ceased publication.

CASE IV

Deana is new at the science reference desk; new, in fact, to reference work, having just recently graduated from library school. When she is confronted by a student who is unable to identify a journal referred to by the abbreviation *Fed. Proc.*, she consults the library's serials list, but finds nothing that matches. Before concluding that the library does not subscribe to that periodical, she decides to identify the full title using the bibliographic sources that are kept at the reference desk.

The title *Federation Proceedings. Federation of American Societies for Experimental Biology* (abbreviated *Fed. Proc. Fed. Am. Soc. Exp. Biol.*) listed in *Chemical Abstracts Service Source Index* (CASSI, Ref. 6.1), appears to match. Deana, having already carefully studied that important reference work, is aware that the AACR entry (which is the entry her library would use) is given in capital letters in CASSI. The AACR entry is *Federation of American Societies for Experimental Biology. Federation Proceedings. This* title does appear in the library's serials list, and Deana directs the student to its location on the shelves. The student drops by the reference desk a few minutes later, journal volume in hand, to inform Deana that indeed she had directed him to the appropriate place.

CASE V

Darin, a senior chemistry major in a small liberal arts college, is planning to go to graduate school. He has several universities in mind for his applications, but before sending them, he wishes to find out more about the research being conducted at each of the schools. By looking into the American Chemical Society's *Directory of Graduate Research* (Ref. 4.2), he is able to find information about the research interests of the faculty members, as well as lists of their recent publications, and the dissertations of their students.

The Corporate Indexes of *Current Abstracts of Chemistry* (Ref.4.8) can be used to identify papers emanating from chemistry departments at specific universities, but this source is limited primarily to synthetic organic chemistry. The Corporate Indexes of *Science Citation Index* (Ref. 22.4) do not have that limitation, and can be used in essentially the same way. For example, Darin might look under "Princeton Univ. Dept. Chem" where there is a list of authors along with journal, volume, page, and year for each. Only the first named author is listed, and that person is very often a student or research associate, rather than a regular faculty member. If the Source Index is then consulted under an author's name, the full title of the article can be found.

The *Directory of Graduate Research* provides more detailed information than either of the other two sources just mentioned, but it is published only every other year, and hence is less up to date.

CASE VI

Shelley is a forensic chemist working for the police department in a large city. She has been given a sample of motor oil and asked to perform analyses on it that will help the police identify the car that it came from. *Standard Methods of Chemical Analysis* (Ref. 15.24) has a long section on the analysis of petroleum and petroleum products, but it does not have the specific information she needs.

When Shelley looks up "motor oils" in the Index Guide to the Ninth Collective Index of *Chemical Abstracts* she is told to "See Lubricating oils." She also looks up "crime," and is directed to "legal chemistry and medicine." In the General Subject Index for the Ninth Collective Index, there are several pages of entries under each of these headings. Although several of the entries under "Lubricating oils" look promising, Shelley's attention is arrested by the modification "automobile lubricants and fuels and soots detection in, by fluorescence" under "Legal chemistry and medicine." The abstract indicates to Shelley that this article (the second one in a series) would be very useful. From this particular article, published in the *Journal of the Forensic Science Society*, Shelley should also be able to locate other relevant publications, including the first article in the series.

CASE VII

Immediately after receiving his Ph.D. in chemistry, Kenneth accepted a post-doctoral appointment to do research at a distant university. His research will require him to make use of Mössbauer spectroscopy, a technique which he is not very familiar with. Kenneth has found *Techniques of Chemistry* (Ref. 15.6) very helpful in connection with other techniques that he has used in his research as a graduate student. He is, therefore, very pleased to find a chapter on Mössbauer spectroscopy in Volume 1, Part 3D (Optical, Spectroscopic, and Radioactivity Methods) of that set. It is very useful, but of course, not sufficient or quite up to date, and Kenneth is particularly interested in reviewing the recent research literature before beginning his appointment.

His new research advisor has suggested that he use the *Mössbauer Effect Data Index* (Ref. 4.17) to locate some relevant literature. This comprehensive source makes it possible for Kenneth to survey the literature without having to search *Chemical Abstracts*. However, the volumes of the *Mössbauer Effect Data Index* usually appear about two years after the publications covered, so Kenneth will either have to rely on reprints supplied by his advisor or on more up-to-date library sources to cover the most recent literature. The *Mössbauer Effect Data Index* is arranged so that Kenneth can identify reviews, theoretical papers, etc., as well as papers that deal specifically with the class of substances he will be studying in his research.

After having mastered a substantial segment of the literature of the field, Kenneth will undoubtedly familiarize himself with the way *Chemical Abstracts* handles both Mössbauer spectroscopy (he should look under Moessbauer effect in the General Subject Index) and the specific chemical substances he is working with.

CASE VIII

In the course of her research, graduate student Victoria is interested in the properties of and synthetic routes to compounds having ring structure C. Her advisor recalls that there was a paper published around 1975 which reported compounds based on that structure, and he recommends that she look in *Chemical Abstracts*. Victoria is not sure how this particular ring system is named in *Chemical Abstracts*, so she looks first in the Ring System Index (Ref. 3.1.5). Note that the Formula Index would be of no avail in this case, because Victoria is not sure which (if any) substituents there might be on the rings.

The structure is a 5,5,6 ring system, with skeletal formula $C_3NS-C_4O-C_6$. At the time Victoria is working on this problem, the Ninth Collective Index is not complete and lacks a Ring System Index, so she

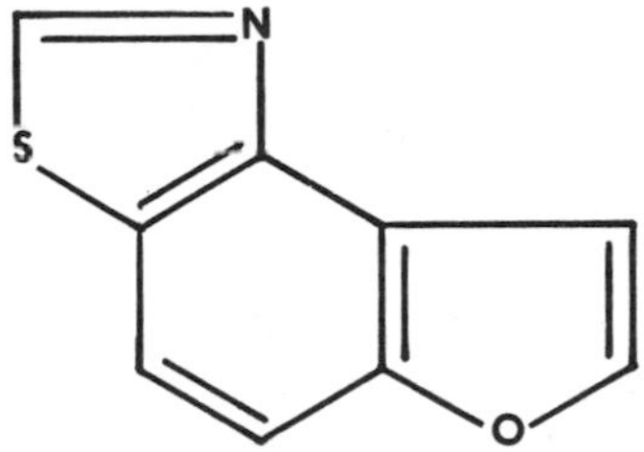

Figure C 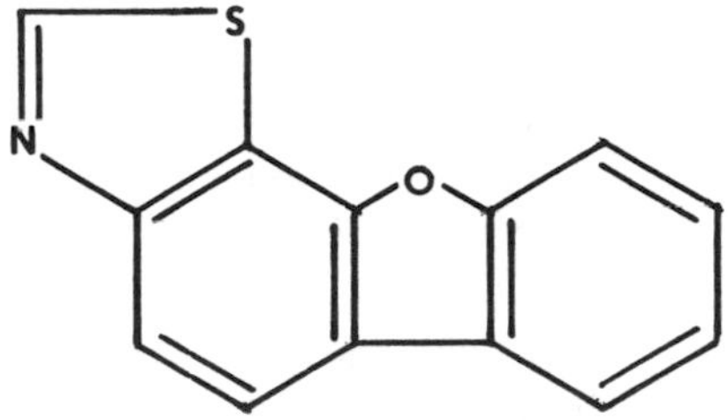   Figure D

looks into the volumes that cover the year 1975 — specifically Volumes 82 and 83, which were published in 1975, and Volumes 84 and 85, which were published in 1976. In Volume 83, she finds an entry for the appropriate ring system, with one name listed — "Furo [3,2,-e] benzo-thiazole." When she looks that name up in the Chemical Substance Index (Volume 83), she is able to confirm it is the structure she is interested in (from the structure drawn there). Three derivatives based on this parent structure are listed, all pointing to the same abstract. Victoria is now able to find and read the paper that her advisor suggested, and may also use the name to search in the Chemical Substance Indexes for the ninth and tenth collective periods to attempt to locate additional references to the ring structure of interest. If Victoria limits herself to this name, however, she may miss some relevant references. In particular, she will miss references to compounds for which the ring structure of interest is embedded in larger ring structures.

Chemical Abstracts Service has recently published a reference tool that can help Victoria with the last-named problem. That tool is the *Parent Compound Handbook* (Ref. 3.7), the appropriate section to look in being the Ring Substructure Index. When Victoria looks there for C_3NS, she finds a number of ring systems which also contain C_4O and C_6 rings. She then must look up each name she is unsure about, by the code given in the index, in the Parent Compound File. None of them meets her specifications precisely, although there are a few, such as Benzofuro [3,2-g] benzothiazole (compound D) which are close. Victoria is also able to ascertain from the *Parent Compound Handbook* that there was not a different name for the ring system in the Eighth Collective Index.

For the seventh collective period, Victoria uses the Index of Ring Systems in the Seventh Collective Index, and for earlier years, she uses the Ring Index (Ref. 3.8). These sources do not reveal any earlier literature on the ring system.

CASE IX

Christopher is a graduate student in geology. He has been studying the magnetic behavior of magnetite under stress, as a technique in earthquake prediction. He is confident that he has already thoroughly covered the relevant geology literature, but wishes to learn if there has been any related research reported in the physics or chemistry literature. After consulting with the chemistry reference librarian, Christopher opts to have a computer search done on his topic. It is a multifaceted topic, for which the CA CONDENSATES/CASIA (Ref. 5.1 and Ref. 5.4) and INSPEC (Ref. 5.7) databases can be searched. Four facets are identified: magnetite, magnetic behavior, stress, and earthquake prediction. Christopher decides that he wants references regardless of whether earthquake prediction is discussed in them, so the last facet is left out of the search.

Magnetite, a mineral, is a specific chemical substance, and hence can be searched in the chemistry files by its registry number. The registry number, 1309-38-2, is obtained by looking up "magnetite" in the Chemical Substance Index, or it is obtained on-line *via* the CHEM-NAME file (see Ref. 5.4). The registry number is of no use for the INSPEC search, so the name "magnetite" as well as its chemical formula is used. It is necessary for the searcher to be familiar with the way various notations such as chemical formulas are handled by the different databases and search services. For example, for an INSPEC search using DIALOG, the formula Fe_3O_4 is entered as

$$FE(W)SUB(W)3(W)O(W)SUB(W)4.$$

For files 3 and 4 on DIALOG (see Ref. 5.4) an algorithm has been applied at the DIALOG input stage which fragments chemical names into chemically significant pieces. Hence, if Christopher's search is performed on those databases using the word "magnetite" (in addition to the appropriate registry number), citations will also be retrieved for the related mineral "titanomagnetite." This may be a bonus, but if Christopher wants to limit his search only to magnetite itself, he may rely on the registry number.

After developing the other two facets with synonyms and appropriate near-synonyms, and formating the search in the DIALOG language, the librarian searches the appropriate databases. Nine citations are retrieved from the CA CONDENSATES/CASIA files and seven from INSPEC. Only one of these sixteen citations is on both of the database outputs.

CASE X

Elizabeth, a chemist in the product development laboratory of a company, needs to know the degree to which dissolving various amounts of common inorganic lithium salts will lower the freezing point of water. Thus, she needs a list of freezing point constants, and/or specific freezing point-concentration data. The index to this guide directs Elizabeth to the Landolt-Börnstein Tables (Ref. 12.11). The annotation given in the guide indicates that freezing point constants are in Volume 3:2a. Elizabeth scans the table of contents in that volume. With the aid of a German-English dictionary (e.g., Ref. 11.14), she identifies the term "gefrierpunktsernicdrigung" as "freezing point depression" and with that term is able to locate the appropriate tables.

CASE XI

Ian, an undergraduate political science student, is writing a term paper on hydrogen as a fuel for automobiles, and wants to include some recent technical information on the production of hydrogen. His chemistry background is limited to two semesters of a university course in chemistry for non-science students. Before attempting to tackle the technical journal literature, Ian reads the relevant encyclopedia articles in *McGraw-Hill Encyclopedia of Science and Technology* (Ref. 22.2) and Kingzett's *Chemical Encyclopedia* (Ref. 10.8). Ian then reads the article on "Hydrogen" in the *Kirk-Othmer Encyclopedia of Chemical Technology* (Ref. 10.17), and although he finds that much of it is of little interest to his needs, and some of it is beyond his comprehension, with his limited chemical background, he is able to glean some relevant information from it for his term paper.

Ian then attempts to locate recent articles on hydrogen production by looking up "hydrogen" in the Keyword Index of the latest issues of *Chemical Abstracts* on the library shelves. The first reference he finds is to a review article in a Japanese periodical that the library doesn't own. Ian doesn't read Japanese, anyway. The second reference is to a four-page report which seems much too specialized to Ian to be of any use in his term paper. The third is to a Dutch patent, the fourth to a theoretical paper which is completely over his head. By the thirteenth reference, Ian has still found nothing in *Chemical Abstracts* that he can make use of in his term paper, although he has spent a great deal of time looking. Partly, this is a case of bad luck. There are some references in *Chemical Abstracts* that Ian can use, but these are vastly outnumbered by those he can't use.

A reference librarian advises Ian to use the *Reader's Guide to Periodical Literature* (Ref. 22.3). There Ian quickly is led to several useful notes in recent issues of *Science News* as well as a fine overview article in the *Bulletin of the Atomic Scientists*. Ian is also able to use some references he locates by looking under "Hydrogen—manufacture" in the *Applied Science and Technology Index* (Ref. 22.1).

Author and Title Index

This index is designed as an aid to locate the entries for specific works known to the reader. First personal authors (or editors) and titles are listed in a single alphabet. All citations are to entry numbers, not page numbers.

Subject Index

This index is intended primarily as a useful guide to the contents of the reference works included in this book. Comprehensive coverage of all the topics in the entries is not attempted—indeed such comprehensive coverage would be virtually impossible. The index can be used to identify one or more reference sources that provide specific kinds of data. The intention is to direct the reader to useful sources—in many cases sources in addition to those listed at an index entry will also include relevant data. Some attempt is made to include references to unusual locations and hard-to-find data. The reader is advised to consider the resources in Chapter 12 in connection with any search for data, at least as good first places to look. Note that, in this index, entry numbers and chapter numbers are given. It is advisable to consult the appropriate annotation in order to understand the context of an index entry.